An Absolute Beginner's Guide to Keeping Backyard Chickens

An Absolute Beginner's Guide to Keeping Backyard CHICKENS

Watch Chicks Grow from Hatchlings to Hens

Jenna Woginrich — *Photography by* **Mars Vilaubi**
Diary entries and creative direction by **Alethea Morrison**

Storey Publishing

The mission of Storey Publishing is to serve our customers by publishing practical information that encourages personal independence in harmony with the environment.

Edited by Deborah Burns and Dale Evva Gelfand
Art direction, book design, and diary entries by Alethea Morrison
Text production by Jennifer Jepson Smith

Cover and interior photography by © Mars Vilaubi, except for © Constance Bannister Corp/Getty Images, 13; Eggantic Industries-www.henspa.com, 33 left; © Adam Mastoon 14, 22, 24–27, 29, 30, 74 middle, 75 (feather patterns); © Drew Waters/Chicken Ark by Handcrafted Coops/Handcrafted Coops.com, 32
Illustrations by © Grady McFerrin

Expert review by Cheryl Barnaba and Maryellen Mahoney
Indexed by Samantha Miller

This book was previously published under the title *Chick Days*.

Storey Publishing
210 MASS MoCA Way
North Adams, MA 01247
www.storey.com

Printed in China through World Print
10 9 8 7 6 5 4 3 2

LIBRARY OF CONGRESS CATALOGING-IN-PUBLICATION DATA ON FILE

For Diana, who handed me my first hen – JW

To Alethea and Xavier, who put as much into this book as I did. – MV

CONTENTS

Thanks So Much

JENNA: I'd first like to thank Diana Carlin, who got me all mixed up in this chicken business to begin with. I'd also like to thank my parents, Pat and Jack, who may be the two most beautifully understanding and patient people a farm gal could ever know. Also, my brother, John, and sister, Kate, who are always willing to hear another chicken story.

I must surely thank Deborah Burns, my Storey editor, as well as Carleen Madigan, Dan Reynolds, Amy Greeman, Deborah Balmuth, and everyone else at Storey Publishing for their partnership in this book and books past.

Thank you to Doug and Nancy at the Wayside Country Store! Thank you to my neighbors Katie, Nancy, Doug, Allan, Suzanne, and Roy, who watch over me and make me feel part of our hollow. Thank you to James Daley, Phil Bibens, Noreen Davis, Paul Fersen, Eric Weisledder, and Tim Bronson, Orvis coworkers who helped with everything from loading chickens into the back of my Subaru to editing book proposals.

Always thank you: Kevin Boyle, Erin Griffiths, Raven Pray Bishop, Sara and Tim Mack, Leif Fairfield, and Shellee and Zach. Thank you to Mary Ellen, my landlord, who turned the key to let this place happen. I'd also like to thank everyone who became a part of Cold Antler Farm's thriving online community and continue to push me to write and keep up with my own dreams every day.

And of course, very special thanks to Mars, Alethea, and Xavier, who shared their joy of new chickenhood with us in their photos and let us tag along as their own small flock grew up.

MARS: To Deb Burns, whose vision launched this book and nurtured it along the way; Pam Art, who tested the book in the field and was always there with encouragement; Dan Reynolds, whose enthusiasm is infectious; Deborah Balmuth and the whole editorial department at Storey for their constant support; Jenna, who inspired us to have chickens in the first place and who found just the right words for this book; Ilona Sherratt, for being the world's biggest chicken lover; Al Whitney, the best neighbor a chicken could have; Maryellen Mahoney, for being our chicken mentor; Gail Damerow, for writing our chicken bible.

Most especially, to Amelia, Honey, and Tilda, for being so patient with their paparazzi!

PREFACE

Welcome to the flock. We're going to visually experience the art of chicken raising by following three special Yankee birds from hatchlings to laying hens. Here in southern New England one young couple and their son decided to take on a whole new life, and they invited some chicks — and us — along for the ride.

After relocating from San Francisco to the Berkshires in western Massachusetts, Mars, Alethea, and their son, Xavier, soon got the chicken-raising bug and purchased three spunky day-old chicks from a local feed store. Without any prior poultry experience, this young family welcomed chickens into their stylish home (literally: a brooder was set up in the spare bedroom), making for some extremely local eggs. Thanks to Mars's photography, we'll watch the chicks' entire life cycle unfold — tailing them from their first days in their brooder to their first laid eggs half a year later.

If you're considering starting a backyard flock, follow along as the laying hens in this book grow up page by page and photograph by beautiful photograph. If you already have a pile of chirping babes under a heat lamp and are nervous as can be, *relax;* you'll be fine — and welcome to the club. And if you grew up with these fine animals and simply want to enjoy the experience all over again, welcome!

Look for Alethea's tips and insights throughout the book, in "Chick Diary" entries that look like this.

Chick Diary

Consider these three girls mentors that will show you the ropes when you order your first batch from the hatchery and then help you better care for your chicks when they arrive. Pull this book off the shelf to monitor and understand your own birds day by day. If you already have a very loud cardboard box in the bathroom and need to know what the heck to do next, we've got *your* back, too. Regardless of where you're at in your chicken-raising dreams, these three hens will help you understand just how easy chickens are by seeing their development right here on the page and knowing what's in store for you over the coming weeks of your chicks' lives.

Jenna, Cold Antler Farm

chapter 1

First Things First: Why Chickens

Why Have Chickens?

Why raise chickens? Because they're the pet that makes you breakfast. They bring home good food *and* belly laughs. Chickens are quirky, beautiful, and oddly clever. They come in countless colors, shapes, and varieties, and there's not a culture on the planet that doesn't raise them. These hardy birds will teach you basic livestock handling and amaze you with their individual character traits. More good news: They don't break the bank. A handful of chicks will cost less to purchase than a large pizza and require less effort than your house cat.

You in so far? Good.

Another reason to raise chickens is the quality of your own free-range eggs, which will bowl you over. No more watery whites and pale yolks. You are in for the richness of a country hen's egg — eggs scientifically proven to be lower in cholesterol and higher in omega-3 fatty acids, keeping you and yours healthier with every new arrival in the nest box. Not to mention these eggs will improve your lovely baked goods and make your omelets tastier.

And my favorite reason to raise chickens: They add life and vigor to your home, turning houses into homesteads and children into naturalists. Pouring scratch grains into a metal bin, closing the coop door at night, mending a hole in the fence so the fox stays at bay — these actions connect us to our food and to our past. Trust me. It's a better life that comes with morning clucks.

Contrary to popular belief, you don't need to live down a country road to keep chickens. Even if you live on the corner of a four-way stop in Portland, given proper care and a little room to flap their wings these gals can adapt and thrive in any environment. What you *do* need is a little bit of space, some research, and a city ordinance that allows laying hens.

Besides feeding you breakfast, chickens are always good for a laugh.

Turns out this isn't asking too much because nowadays people are keeping chickens in places no one considers cliché. Young couples in suburbia have Ameraucanas perching on flowerpots and kids racing past Wyandottes when they fly out the back door to jump into the car for football practice. They're keeping these birds because they want to know where their food comes from, sure, but they're also keeping them because having chickens is fun and easy, and it's hard to be bored mowing your lawn when a trio of hens is waddling behind you for the free salad bar.

COMING HOME TO CHICKENS

I grew up in a small town in northeastern Pennsylvania where the keeping of livestock was verboten. Oddly enough, my mother grew up in that same town in the 1960s — and every house on her childhood block had a coop in the backyard. But somewhere along the years the town of Palmerton changed its zoning, forbidding farm animals within city limits.

I blame microwave ovens. Modernization and gadgetry made chickens seem either lower class or too country, so the birds were turned into contraband in some sections of Carbon County. After all, you don't move to cultural epicenters like Palmerton (population 5,200) to see the likes of scrappy yard birds. (That said, New York City never felt the need to change its pro-fowl laws in all five boroughs — including Manhattan. . . .)

So, chickens being outlawed in my hometown — and totally out of the question, anyway, as far as my mother was concerned — they never entered my mind. I went off to college (leaving home for good, it turns out), graduated, and was offered my first bona fide design job down in Tennessee. I lived the city life for a while, but I knew in my heart that lifestyle wasn't for me. Within two years I was packing up my station wagon again, this time heading to a former cattle farm in northern Idaho.

My new job and the farmhouse I rented were both pretty chicken friendly. That was all the coaxing I needed. Somehow the stars aligned, and I discovered Diana, a coworker who had a two-hundred-layer natural-egg business. With her kind help and some trial and error in Backyard Flockmanship 101, I learned the ropes. It took all of four days to realize I was hooked. I never once regretted the decision. I can't say that about many things.

I started keeping chickens only four years ago. I've kept them in pens, hutches, "recycled"

Chickens used to be common in backyards everywhere.

coops (a fancy way of saying a compilation of stuff from the dump), and on both coasts of this fine nation. In that short span of time, my life's orbit has changed, based on keeping poultry, so any place I'll call home will need to be egg-production friendly. The idea of a life without pullets in the spring seems depressing at best. Chickens have a way of taking you to another place. Something about a red hen bobbing her head past a kitchen window instills a familiar, if long-buried, comfort — a sense of home.

Chickens stand for a simpler life — a source of yummy natural food, sure, but actually having some in the garage changes you. For me they opened the door to organic gardening, sheepdog trials, pack goats, and canning in my August kitchen. I will never go back to that life before *pullet* was part of my vocabulary, and I don't understand how anyone living within the right zoning codes can. They're as easy to tend as a stray cat, and the time needed to take care of them daily amounts to what most people waste waiting in line for coffee. They're funny and have their own individually quirky personalities that can't help but grow on you. After a few springs of hatchery orders, I'm a convert turned preacher.

PART OF THE SOLUTION

The factory-farm norm behind supermarket eggs is a sad story. Confined chickens live out their lives at best in indoor barns that double as feedlots — or at worst in tiny wire cages. They live in a hell of stress and misery, and the end product is a shadow of the once-great egg. Factory chickens produce thin-shelled, yellow-yolked, slimy protein glop compared to the robust and hearty flavor of a farm-fresh egg.

When you keep your own chickens, you are choosing to walk away from the factory farms with their inhumane cages and amputated beaks. You're taking back a little freedom — both yours and the chickens' — and it is delicious. In more ways than one.

Home-raised hens produce beautiful eggs in every color of the sepia-tone rainbow.

If We Are What We Eat, Let's Be Something Better

FOR YEARS WE AMERICANS HAVE CHOSEN cheap, seasonless selections and endless variety over healthy, in-season crops raised without pesticides. Some say that a local organic diet is an elitist goal that regular folks can't afford. We've bought the lie that eating whatever we want of lesser quality is a good thing because it's easier. And because we don't have to connect the cow with the burger — or the caged hen with the egg.

Ask the average American if he or she would rather buy a feedlot chicken pumped full of antibiotics yet riddled with bacteria or drive to the farmers' market down the block and pay a dollar more a pound for a free-range, disease-free bird, and most would prefer the healthier option. But oddly, few choose it.

And the misconceptions are rife. One organic farmer was almost shut down for processing his poultry outdoors near the fields they pastured on — the rationale was that outdoor living rendered the birds unclean. So he sent a large sampling of his stock and an equal sampling from a grocery store to be tested for bacteria. His came back ridiculously healthier — without the benefit of multiple chlorine baths and a "safe" packaging plant.

I understand that we have a world to feed. The point is not to boycott the grocery store but to change what's inside. Buy local, buy organic. Ask the clerks what was grown in your area. Show the people who order produce that this is what buyers want, and things will change.

While not everyone can afford a steady grocery-store organic diet, most of us can afford one local meal a day. Experts say if once a week every American ate a meal that was produced within one hundred miles of his or her home, the food industry would be forced to change dramatically. Organic wouldn't bust the budget, it would be normal. Start however you can. Get some cheap local oats for oatmeal at the farmers' market, and you've just eaten a breakfast that can change the world.

Unfortunately, most people don't want to think about where their food comes from. They don't care about local farmers. They don't want to buy healthier meat for more money and eat less of it. They don't process how recalls of poisoned peanut butter and salmonella outbreaks relate to their buying habits. Life is complicated enough and hard enough without having to think about where to buy food. But it really does come down to consuming meat, eggs, and vegetables that won't make you sick — and that also won't consume natural resources. If we are what we eat, let's be something better.

A New Edible World

Get a few layers in the garden, and things gradually change. You'll notice your ideas about food evolving as you progress from consumer to producer. It gets harder to pull up to drive-thru windows and easier to cook at home. Food-shopping trips lead you to farmers' markets instead of fluorescent-lit grocery aisles. Opt out of the normal route to eggs, and suddenly other paths to homegrown and homemade foods reveal themselves to you. Once I began collecting eggs from nests, I learned to bake piecrust, make cheese, and grow my own pizza toppings in my organic garden — a feat I would have never considered before homegrown scrambled eggs were on the plate first.

Once you start picking up eggs off the lawn, you let the food chain into your life — and you start wanting more and more. Soon tomato plants are yielding fruit where frilly annuals used to bloom, raspberry plants supplant the perennial garden, and simmering tomato sauce on the stove and canning homemade jam become the norm. You might start reading about dairy goats or learn to churn your own butter.

This may sound slightly ridiculous, but just wait and see. Chickens make any homemade food adventure seem possible. Once you see how easy tending them is, you'll start planning for next spring's beehive or next fall's apple trees. Rabbits could be next, or ducks. You'll have crossed over to the farming side of the road. Sustainability is the best kind of addiction — one that doesn't require weekly meetings to overcome.

Better start bookmarking those recipe sites now. You'll be hooked.

Chickens are the gateway to a host of other livestock, such as rabbits and ducks.

My first postcollege job was in Knoxville, Tennessee. I moved there by myself to work for a television network's website. I rented the bottom floor of an old boardinghouse in a historic district called Fourth & Gill. I'm pretty sure that old place could fit two of my present cabins inside it. Maybe three. It feels like ages ago. A past life.

Back then all I wanted was to be a designer. I wanted a board position in my AIGA chapter. Little did I know 18 months later I'd be in a farmhouse in northern Idaho or that I'd soon relocate again to a New England hollow and a small cabin on a patch of land I call Cold Antler Farm.

If you dream of goats and chickens and a cabin in the woods but are presently sifting through take-out menus in your current metropolis, please remember that just a few years ago I had one dog in a city apartment. Now I'm in this beautiful mess.

Tomorrow I'll visit a brewery and probably come home wanting to make my own beer. Sunday, Steve and I are going to slaughter an angry rooster I raised out of the palm of my hand. Right now I'm going to go outside and close the coop door before the rain comes.

If you wish you, too, were closing a coop door, I promise if it's something you really want, it'll happen. You'll find a way because you must. And when it does happen, be ready because it'll come fast. Life doesn't happen any other way. At least not the parts worth living.

Gossip, Rumors, and Big Fat Lies

I feel obligated to give a friendly warning: when you start telling friends and coworkers about your new livestock adventures, skeptics of backyard chickens (and you'll meet quite a few) will try to be overly helpful. Unfortunately, their idea of "helpful" will be telling you what a horrible idea it is to get chickens. I will only say this once, and I mean it with every fiber of my being: they are wrong.

Some well-meaning but poorly informed people may make heartfelt interventions to save you from a future of bib overalls. They will say the word *chicken* very much like they'd say *sewer rat*. Chickens? That just isn't what normal people do, is it?! They'll wave their hands dramatically and warn you that chickens are some giant commitment. They'll scold you, telling you that hens are a needless expense. They'll tell you that chickens are too eccentric for suburbia or too smelly for your urban 10 × 20 plot. Occasionally someone will raise an eyebrow and ask, "Can't you just *buy* eggs?" and look at you with sincere pity that you'd even consider such a ridiculous notion. Some kind but clueless souls will be concerned that you can't afford eggs in their prepackaged Styrofoam containers at the store. Bless their hearts.

SEPARATING FACT FROM FICTION

I'd like to take a moment to exonerate the fine hen from some of the other smack talk going on behind her back. For an animal as common as the chicken, it constantly amazes me how ignorant the average person is about them.

For example, a woman who wanted birds for years but never went about getting layers explained that roosters weren't allowed within her town's limits due to the noise issue. (We would agree that is a bummer but

Hen: An adult female chicken

Rooster: An adult male chicken, sometimes called a cock

Pullet: A female chicken under one year old

Cockerel: A young rooster under one year old

Bantam: A miniature chicken, sometimes called a banty

Layer: A hen suitable for egg production

Broiler: A chicken (of either sex) suitable for meat production

Vent: The all-purpose exit chute on your hens' bums

Molt: The annual shedding of feathers; birds don't lay when molting

Sexed Chicks: Chicks separated into pullets and cockerels

Straight-Run Chicks: Nonsexed; a variety of pullets and cockerels

understandable). She went on to explain that she wanted chickens for eggs, not meat, and without a rooster they wouldn't lay, so it was pointless.

In fact, roosters don't make eggs happen; the egg is just part of the hen's natural cycle — just like a woman's. All females carry eggs in some form or other and pass them as part of life. A hen just does it every 25 to 36 hours instead of every 28 days. No one on this side of the gender fence has ever needed a fella to make an egg happen.

Another common misconception is that chickens are loud. Truth is, most hens rarely make a peep. They can be vocal, but it's nothing like the piercing cries of geese (I have a pair louder than any car alarm) or a barking dog. Chickens do cluck and coo, but it's the noise equivalent of leaving the classical station on low volume in your backyard.

And as for that "big investment" rumor: Sure, you can buy designer coops and order expensively bred show chickens. But like any hobby chicken raising can be as frugal as you choose. A practical beginner without any resources or carpentry skills can easily purchase and set up an egg shop for under $350. That's a small investment for years of healthy, local protein.

The biggest fib would be the time you'll need to dedicate to your omelet shop. Yes, you'll have to check in on and feed the layers daily. My routine: Before I head off to work, I open their coop door, throw down some grain, and make sure they have fresh water. I do this with a hot mug of coffee in my hand and a grin on my face and it never takes more than 5 or 10 minutes. At night it's slightly more effort, but only because that's when I collect eggs and refill the feeder. Total: maybe 15 minutes. The average person is willing to stand in line for a movie longer than that. And half the time the movie stinks. I see a clear winner here, don't you?

Geese are noisy.

Chickens are not.

Taking You Home Again

People have many reasons for keeping chickens. Some are being practical about healthy, cheap food while others are striving to be as self-reliant as possible. But I think there's something to this pull other than frugal living and good French toast: a panting-dog desire to go back to simpler times.

HOMESTEADING PETS

Imagine the satisfaction of scrambling eggs that you collected from your own flock for your brunch guests. And you just can't help but cross your arms and smile when you walk out on the back deck at dusk in late summer and see your pets providing equal parts pest control, food production, and entertainment. When I look outside at my yard, I see not just chickens but the beginning of a journey dedicated to sustainability and an avatar for a better, simpler life. They truly are the homestead pet for everyone.

Keeping chickens frees up our spirit. I don't mean to sound overly corny — I really am a grounded gal — but knowing that I can have empty cupboards and no money to my name and still eat an amazing breakfast of fresh eggs without ever leaving the cabin . . . well, it makes me feel rich. Not some superficial wealth but comfort in knowing I'm okay. These girls have my back.

START COLLECTING THOSE CARTONS NOW

While chickens also make for a great introduction to meat farming, this book concentrates on laying hens. Healthy, natural eggs are the focus here. A small, perfect food containing all the building blocks of life because, quite frankly, it *is* the building block of life. Eggs are powerhouses of protein and energy. The omega-3 fatty acids are proven to help everything from the thickness of our hair to the balance of our mood.

Collecting these small miracles right at home is what motivates most folks to keep a few birds under the trellis. The same oval you've cracked and eaten your whole life suddenly goes from the world of consumer to that of producer. Eventually you'll walk by the cooler in the grocery store, and you won't be able to wipe that smirk off your face. *Buy eggs?!* Please. That's . . . well, that's just ridiculous.

chapter 2

Next: Which Chickens

A Breed for Everyone

When choosing hens for your flock, it helps to know a little about the history and characteristics of each breed. You'll learn which are better suited for snow and indoor confinement and which need open spaces. You'll find out which will be likely to sit on eggs and go broody and which would never qualify as parental stars. (See also the chart on pages 114 to 119.)

Some birds do great outdoors and require little to no feed if they can scavenge on their own. Some gals lay like champs in close quarters while others need plenty of free-range space to spread their wings. Some lay right through a cold winter, and others shut down when the days grow short. And as for temperament, some are major characters while others are calm and gentle. On the flip side, giant corn-chowing roasters developed for feedlot barns should never be considered for laying stock since most won't survive much longer than the time it takes for them to reach slaughter weight.

There are egg-color preference and noise levels for you to consider, as well. All of this should go into your decision of what chickens are right for you and your home. So do your homework!

Finally, remember that you can't have just one chicken. These social, flock-living animals need the company of at least one other hen to be content. Best is a minimum of three. The birds can be happy as a flock on their own, and if one passes away for any reason, a social bond is maintained.

Silkie bantam

Australorp hen

My Top-Ten List

Here is a list of ten popular backyard birds that are proven layers and/or just plain fun with good temperaments. I've personally raised them all over the years, coming to know them not just as livestock but also as individuals. (It is amazing how different a young Rhode Island Red is from an older Brahma.) If any of these sound interesting, check out some breeders online, and ask questions. You'll be glad you did the research, and it will certainly help to ensure that you're set up for success.

SILKIE BANTAMS

THIS BREED FROM JAPAN was given the name Silkie to describe its unique feathers. Unlike other breeds of chickens, these beautiful oddities have soft, flowing, furlike feathers that remind me more of a shih tzu than a chicken. Their funky look, smaller size, and gentle demeanor make them excellent starter birds. They'll show you that real chickens can look a lot more fabulous than what decorates your average farmstead needlepoint.

Silkies were my first chickens, and they were such a great experience, I have to suggest them. While they are bantams, and the eggs are smaller than normal standard-layer fare, what they lack in size they make up for with personality and mothering skills. Silkie Bantams are amazing brooders and caring mothers. Some poultry farmers slide other hens' eggs under these mamas to ensure proper parenting. They're also incredibly calm around people and animals, making them a great bird for families with young children. And of course, there are the crazy looks of these birds: all poof and feathered feet. They do well in confinement and in many climates. An absolute A+ chicken.

AUSTRALORP

AUSTRALORPS ARE AN AUSTRALIAN INVENTION. Hailing from a heritage of Rhode Island Reds and Black Orpingtons, along with a few other high-production breeds, these black birds are serious about laying and have a reputation for high production and sweet temperaments. They're also a good choice if you want a dual-purpose bird: one equally great for meat and eggs. If you're thinking about this breed, it's a good idea to order a straight run — meaning not sexed — of these chicks. You could keep the ladies for their eggs and dispatch the roosters once they hit table weight.

Another big perk is these gals' noise level. Australorps are quiet birds, and if you have neighbors close by and still want fresh brown eggs, they may be a good match for your lifestyle. They're also fine with confinement, making them suitable for a backyard pen. The only downside is these shy birds may become easily dominated by more gregarious breeds if they share a coop. Barred Rocks or Rhode Islands may boss them around. They're a good sister bird for Brahmas and Orpingtons, though.

Meat or Eggs . . . or Both?

WHILE MOST PEOPLE KEEP A SMALL FLOCK FOR THE EGGS, you might want to try your hand at raising chickens for the roasting pan, too. If you aren't one to get overly sentimental about your chickens, add a few broilers to your spring hen order. But maybe you feel that killing a chicken for its meat — especially one you've named — is no different from killing and eating your cat. Really, it all comes down to what these birds are to you. Are they pets? Are they groceries? Or are they both?

On the other hand, what happens when hens get too old to produce eggs regularly? Sure, they can live out their years on their roost till they peacefully fall into the straw. Or you can make the best chicken soup you ever tasted by dispatching them as they grow too old to be reliable producers. There is no better-tasting, healthier, and more humane way to be a carnivore than to raise your own meat.

If the idea of raising animals for food doesn't bother you, but doing them in yourself does, you have options. You can transport live birds to local small butchers who will process them for you — you just write a check. Or perhaps a friend or neighbor would "dress" the birds for a small fee or a trade. If you do order some Cornish roasters, raise them along with the layers. In a few months you'll not only have all the scrambled eggs you can eat, you'll also have roasting birds for your Sunday oven.

WYANDOTTE (Silver-Laced): *A favorite among backyard chicken keepers, the Wyandotte is a dependable choice for both eggs and meat. Unlike many other breeds, a dressed Wyandotte will look similar in proportion to the chicken meat you buy at the store and may be appealing for that reason.*

CORNISH (White): *This is the bird you'll find in the supermarket. Conveniently for the meat industry, Cornish hens and roosters are almost the same in shape and size. This breed is really just for meat or show and will reach butchering weight quickly (eight to ten weeks). The hens are poor layers.*

SUSSEX (Speckled): *Before breeders developed the modern Cornish, the Sussex dominated the English market for meat and eggs for almost a century. Like the Wyandotte, the Sussex is both a prolific egg layer and a heavy bird perfect for the table.*

BRAHMA

LIKE THE FAMED BRAHMA CATTLE native to India, Brahma chickens are big and gentle. These large birds (hens weigh nearly 10 pounds!) have feathered feet, small combs, and big hearts. They come in several colors, with buff, light, and dark being the most common. These easy-to-care-for hens won't be carrying on at five o'clock in the morning, they do marvelously in colder climates, and they get along swimmingly with other breeds. Though happier as free-range chickens, they can make do in confinement if given enough legroom. They go broody fairly easily, but sometimes eggs crack under their weight. Hey, nobody's perfect.

However, it's the attitude that wins me over with these chickens. If a chicken could be in the running for a Buddhist monk, these old-time gals would already be ordained. I've owned a few over the years, and hands down they are the calmest in the coop. Though not anywhere near industrial-level layers, they are steadfast, producing even as the temperatures drop way below zero. If you live in a colder climate and want the draft horse of the chicken set for a free-range flock, these are your girls.

Brahma hen

PLYMOUTH/ BARRED ROCKS

DEVELOPED IN NEW ENGLAND, these brown-egg layers are great additions to a backyard flock. Plymouth Rocks are usually black and white (barred), but they now are bred in a variety of colors. The "every chicken," combining the traits of many breeds for many purposes, they are the poster bird for backyard poultry since their production, noise level, and temperament are right in the middle of the spectrum.

Adaptability is this dual-purpose bird's greatest trait. They can hang out just as happily in a fenced pen as they can free-ranging outdoors. There are no surprises with this breed; they are flat-out regular chickens and like it that way.

Plymouth Rock rooster

RHODE ISLAND RED

AN ALL-AMERICAN CLASSIC. From the state that bears their name, these girls have become farm heroes and free-range superstars. A dual-purpose breed, their high-volume brown-egg laying traits and meaty frames make them a perfect hobby-farm starter bird. They're a ruddy brown with black tails and unlike many of their contemporaries have been bred more for work than for show. Rhode Island Reds are the border collies of the laying-hen world: a little scrappy but completely utilitarian. Roosters can get aggressive, but in their defense, they are extremely protective of their flocks.

Hardy and bursting with character and curiosity, they are fairly vocal birds, always the loudest in my flock. When a Red lays an egg, you'll hear about it — even across the yard and through a kitchen window. While the hens are manageable, they aren't docile by any means. They're tiny dinosaurs, always foraging and on the hunt for a stray caterpillar outside the garden fence.

Rhode Island Reds are rarely broody, so don't depend on them to hatch future chickens. However, what they lack in domestication they make up for tenfold in consistent egg production. They're a great, useful, and colorful character on the farm, which is where these birds belong. They can handle confinement and small lots, but they truly shine on open pasture and old fence posts.

AMERAUCANA

THIS BREED COMES BY WAY OF SOUTH AMERICAN PARENT STOCK imported in the 1930s called the Araucana. In the 1970s American fanciers started developing the bird that many of us know and love today, so the name was adapted to its new Yankee home: Ameraucana. Hens are midsize with beautiful, variegated plumage and slate or green feet. They have puffy cheeks ("muffs") and large eyes that remind you more of birds of prey than a chicken. Ameraucanas also have a coo and a cluck unique to their breed. They seem to hum and trill while the other birds around them squawk and clatter. Get a motley crew together and you'll see what I mean: those girls' songs really stand out! They are a lot like Rhode Island Reds in their metabolism and activity level but far milder in noise levels and calmer in your hands.

The real show-pony feature on this bird isn't its voice or feathers but its eggs. Ameraucanas have forever left their mark on the cottage-egg industry because these little girls lay the famous Easter eggs: blue, green, and pink. To come out to the coop and see a small, blue jewel between the cream, brown, and chocolate-colored eggs is a treat I never get tired of. Customers (once they get over the mild shock and realize how tasty they are) seem to covet them as well. When I deliver a dozen eggs to friends and they see their green or blue egg, they ask questions they didn't realize they even wanted to know about chickens. I don't think my farm's Ameraucana intended to do any PR work in her daily laying, but she sure has.

Rhode Island Red hen at left and Jersey Giant hen at right

Ameraucana rooster

JERSEY GIANT

FROM THE GARDEN STATE ITSELF, the Jersey Giant is a contented bird in any coop setting. Just keep in mind that these are big girls. Real big. See the photo on page 27 for a visual comparison to a Rhode Island Red, which is a fairly average-size chicken. Jersey Giant hens are easily over 10 pounds, and roosters can outstride small children. If your coop space is limited, you can easily get four Ameraucanas in the space needed by two of these girls. But if you have a free-ranging flock, they'll make a great home team.

Like many of the heavy breeds, they're calm and docile. Jersey Giants are easy to handle and fairly quiet. (In fact, I don't think I've ever heard any of mine ever make a sound — but I don't think there's such a thing as a silent chicken, so I'm sure it's a matter of timing.) The only real downside is how slowly they mature. It may take up to two years for the plump hens to fill out and reach their adult weight, and that slowness to reach adult size pushed them out of the mainstream when bigger, faster, and cheaper was what farmers wanted. But these girls are dependable layers, and their light brown eggs are a nice contrast in the basket next to the darker browns of most other production birds.

ORPINGTONS

HAILING FROM KENT, ENGLAND, these chickens were bred into what we know as the modern Orpington back in the late 1800s. Developed from a lineage of Plymouth Rocks, Minorcas, and Langshans, these dual-purpose birds were bred to perfection. The outcome: a hardy English farm bird created to withstand cold winters, be easily handled, and like most big gals, keep laying through winter. Orps have a large frame, a broad back, and very soft feathers.

These popular farm birds might also be the perfect urban chicken. If your neighbors are wary about the idea of a henhouse next door, get Orps. It's the chicken equivalent of getting a Golden Retriever to introduce people to dogs. They come in several colors, but the Buff Orpington really is the beauty queen of the lot, with feathers in a creamy orange reminiscent of golden fleece. Quiet, clean, calm, and laying beautiful pinkish light-brown eggs, Orps do well with other heavy breeds like the Australorp, Brahma, and Jersey Giant but fare poorly with more aggressive breeds. I've heard that some mixed flocks pick on these girls the most due to their subordinate nature.

Chicken Colors

FROM PARTRIDGE TO LEMON BLUE, chickens come in dozens of different color and feather patterns. Each photo here shows only one example of either a hen or rooster, but there are usually several color options within a single breed.

The Orpington, for example, can be black, white, buff (a golden color), or blue (a slaty color). Plymouth Rocks come in a variety of feather patterns, such as barred and penciled. For examples of some common feather patterns, see page 75.

DOMINIQUE

THIS BREED, TOO, ORIGINALLY COMES FROM THE UNITED KINGDOM, and it's rumored to be the oldest breed in America. Legend has it that the Pilgrims brought these girls to the New World, and the breed's name is associated with the island of Saint-Domingue (now Haiti). A black-and-white barred chicken whose color resembles the more modern Barred Rock, this breed is old school. It's smaller and less jolly than Rocks — but not to a fault. Its medium size (about 6 pounds) makes it a good match for Ameraucana and Rhode Island Red flock mates, though Doms are certainly the calmest of that trio. Dominiques are a heritage breed great for close quarters; they do well in confined spaces, such as indoor urban henhouses.

The Dominique (along with the Jersey Giant) is the rarest of the backyard chickens I'm listing here, and that's a point that shouldn't go unnoticed. These and other wonderful birds are on the American Livestock Breeds Conservancy's Watch List. By raising a few in your own home you're not only getting delicious future frittatas, you're helping to keep a bit of our history alive. Not a bad deal.

Orpington rooster

Dominique rooster

Chicken Sizes

MINIATURE CHICKENS ARE KNOWN AS BANTAMS. Some breeds, such as the Silkie, are "true bantams" and come only small-sized. Most other breeds, including all of the others in this top-ten list, are bred in standard as well as bantam weights.

Damp conditions can make these minis vulnerable to a variety of health problems. If you're raising banties in areas with a hard winter you'll need to take extra precautions to keep them dry and draft-free during the cold months. In hot climates, make sure to provide plenty of ventilation and a good fan.

WELSUMMER

I AM CERTAIN THAT THE ROOSTER ON THE CORN FLAKES CEREAL BOX IS A WELSUMMER. This bird is the real deal: a beautiful brown hen and a handsome rooster. A medium-size bird, usually partridge brown, it's not a fancy chicken by any means but a truly low-key beauty. These are birds you look at and then do a double take because you didn't initially notice the patterns on the brown feathers or the green sparkling tail feathers of the rooster.

Welsummer rooster

Welsummers come from Holland and are a cross of several breeds, including Barnevelder, Wyandotte, and Rhode Island Red. They aren't super friendly, but don't mistake their aloofness for anger. They're just independent women. They can stand up against the crankiest production bird with gusto, and their dark brown eggs are so amazingly beautiful, you might mistake them for chocolates.

Ducks, Geese, and Turkeys, Too

DEPENDING ON YOUR LIVING SITUATION, you might consider adding other types of poultry to your flock. Ducks, geese, guinea fowl, and even the occasional turkey can be raised alongside chicks. They can eat the same medicated starter feeds as pullets and when young (up to 3 years old) can mingle with the laying hens. Place a gosling in with a pile of Barred Rocks and he'll make himself right at home. Just be sure to keep the bedding clean and dry.

Mixing up your poultry has more to do with your space limitations and town ordinances than it does with the ability of your animals to get along and grow up healthy. While chickens are quiet and inconspicuous, geese are loud — sometimes *very* loud. And some local legislation may allow for three hens but draw the line at a webbed duck's foot. So before you decide to expand your world of poultry, do your homework, and be honest about your limitations.

chapter 3

Hen Housing

Planning Your Coop

Before you commit to bringing chicks into your home, you need to think about housing and consider several basic questions: How much space will your adult chickens need? If you live in a cold climate, what kind of winter housing will you provide for them? And how much time and money are you willing to lay out to ensure that your hens have proper quarters? You should have these essentials worked out before you head down to the hatchery or place your online order.

By the time your chicks start resembling min-. iature chickens instead of those round balls of fluff you carried home from the feed store or post office, you should set up their permanent outdoor housing. It can be humble or lavish, as long as it's functional. The coop will be their safe haven — shelter from storms as well as claws and jaws. It's the only home they'll ever know, and you need to fully understand and respect that responsibility.

OPTIONS

The number of chicks, your housing situation, and how much coin you want to turn over will all factor in. If you rent a brownstone in Memphis with a fenced backyard, you could get away with nothing fancier than a converted doghouse you scored off Freecycle. If you own a half acre of suburbia, you could buy building plans to perfectly suit your birds' needs or order a prefab chicken spa complete with built-in nesting boxes and watering stations. Don't limit your options — get creative!

How creative are we talking here? Plenty of small structures make for perfect chicken coops even if that wasn't their original intention. I've seen old VW buses gutted and turned into chicken coops. I've seen theme coops based on outhouses, post offices, and summer cottages. (My all-time favorite had an old movie-house sign reading "now playing" — with the *p* removed — on the coop door.) I once used an old metal barrel and two apple crates with a shoddy tin roof to make a home for some Silkie bantams. It wasn't as cool as the Volkswagen, but it did the job — and if the tenants had any aesthetic complaints, they kindly kept them to themselves.

BUILD OR BUY?

If you're comfortable around a circular saw and want to call out your inner artist, you can literally build a future for your girls. But even if

This portable chicken coop allows you to move the chickens around the yard. See Buying Notes on page 125 for more information.

you lack carpentry skills, don't think you need to spend a grand on those 4 × 6 coops you see advertised in magazines. Chances are an old shed or a dog run — or even a used coop — is just waiting for you to purchase for cheap in your local paper, Pennysaver, or the Craigslist farm and garden pages. Borrow a friend's pickup or rent a U-Haul for the day, and go get it.

You also have the option to buy a new basic coop, and when you're dealing with fewer than five birds, there are quite a few affordable options. My praiseworthy first coop, the Chik-n-Hutch, cost less than two hundred dollars (with shipping), and I assembled it with only a screwdriver in twenty minutes on my back deck. It made it through an Idaho winter and — covered with an old wool army blanket at night — sheltered my hens against coyotes, wind, and rain. Looking back, a coat of outdoor-furniture spray paint and some nailed-down metal roofing would have made it perfect. If you take such a route, consider doing some simple, cheap renovations before the chicks move in.

As backyard chickens grow in popularity, so do coop solutions. I've seen adaptations of garbage cans and compost turners. I've also seen futuristic space coops that look like those old iMacs. A multitude of options are compiled and cataloged in books and pamphlets. Or check out online sites like Backyardchickens.com. Poultry e-mail list serves, local fancier clubs, and homesteading communities are all available to you at the click of a mouse. These sites are thriving collections of chicken people dying to give you tips and inspiration to help you get started in what they've come to love.

For those raising chickens in defiance of local ordinances, there is the stealth coop, which disguises the henhouse as a trash can. See Buying Notes on page 125 for more information.

An enterprising first-time chicken owner builds a coop of his own design in his garage.

Foxes and Hawks and Coyotes — Oh, My!

Not to mention weasels, bobcats, fishers, owls, raccoons, and various stray dogs and cats. They're all out there, crouching in the hedgerows smacking their lips and scanning for fresh chicken dinner. It's your job as the guardian of the flock to keep your chickens safe so that they can produce for you and yours for years to come.

If you live in a city, you may think that means you won't have to worry about predators — especially the larger ones, such as coyotes — harming your flock. But whether your backyard setup is urban, suburban, or rural, it will attract animals wanting to crib a meal from your egg factory. No hen is ever *really* off the radar, but that doesn't mean keeping chickens is a synonym for homeland security. There are basic precautions you can take, and your options vary.

At a minimum, always lock up your flock at night. I don't mean turn a literal key; rather, make sure predators can't invade the place where your birds tuck their heads under their wings. Even in the most urban of areas, your girls still run the risk of becoming a meal, so your coop should have a latching door not easily flipped open and chicken wire over any windows or openings.

If your coop isn't inside a fenced pen and doesn't have a floor, make sure the base is secure from predators that dig, like foxes and dogs. You can attach a foot or two of chicken wire to the inside base of your henhouse and let it rest on the floor around the entire perimeter. It may discourage digging animals to meet a wall of wire just when they think they're hitting pay dirt.

FREE RANGE OR PEN SAFE?

We all love the notion of our very own free-range backyard eggs, and so do your neighborhood predators. A bird that can roam as it pleases in a safe, fenced yard is certainly a happy bird, but you might want to limit its free-ranging time to when you're home. When an Idaho neighbor complained about my chickens walking onto his property I had to confine them, letting them range only in the evenings while I made dinner. Birds are less likely to stray too far from their home come sunset, and this way I could keep an eye on them from my kitchen window. Compromises can be reached.

Since my time in Idaho, my laying flock has been entirely free range. I currently live in a rural area where locals are used to swerving around the occasional goose or hen. During the day my birds have plenty of cover from aerial predators, and come nightfall they are locked up in their coop, protected from nocturnal beasties looking for a midnight snack. I have lost some birds to a fox, but for the bliss of my roaming birds, the occasional loss is one I'll abide.

If the fox were to take a large quantity of my girls, however, I would certainly confine them to a fenced run for their own safety and my peace of mind. Confinement options include stationary pens, electric portable poultry fences, and mobile pens. Stationary pens, of course, keep your birds in one area, which they will deplete of grass and turn into a dust bowl in no time. Portable fences and mobile pens, on the other hand, let you move your birds daily onto fresh patches of grass, giving your lawn a nitrogen fix and your birds a fresh area to explore without staying long enough to ruin it.

If you don't have time to move birds and want them to be foxproof, then a stationary pen is best. If you have a few minutes to move them across the yard, try the mobile unit!

Talk Is Cheap – and Effective

I HAVE LOST QUITE A FEW FOWL TO A RED FOX at my farm. In retaliation I tried everything, including padlocking the coop, a long stakeout with my .22 at 5 AM, and even a Havahart trap with a turkey breast lure. Well, the fox ignored the lock, avoided my rifle, and tunneled under the trap to eat the turkey breast through the bottom of the cage. I was dealing with a professional. Who would have guessed that talk radio would save the day?

The idea is to use your radio as an audio scarecrow – or in this case, a scare fox. Tune your dial to a stream of human voices, and hang it from a tree or fence post. So far it's been working for me – plus I've finally discovered a suitable purpose for Sean Hannity. Even at the height of what would normally be prime fox feeding time, I've found it's safe to leave the birds alone while the radio's on, feeling confident enough to take a good long walk away from the scene of the potential crime.

GOOD FENCES MAKE GOOD NEIGHBORS — AND SAFE(R) CHICKENS

A strong fence for your stationary pen can be a definite deterrent to predators and offers considerable protection. Even chicken wire several feet high will keep a lot of terrestrial predators away. But your fence may be useless against a burrowing fox unless you take special measures.

To make your girls' quarters more secure against diggers, bury the protective fence or chicken wire about 12 inches deep and toe the fence outward about 6 inches. Even the larger predators generally won't dig that deep under the fence to reach your flock, and the wire toed outward means any excavation attempts will merely result in running into wire. Of course, fences won't stop avian predators like hawks and owls, so. . . .

DISTRACTIONS AND REMINDERS

Predators scare away other predators. Set up one of those inflatable garden owls to make any hovering hawks anxious. I'm not sure how long the local talons will buy the fake owl bit, but I've heard from many chicken owners that they're better than a fence in some cases where foxes are rare but raptors are not. A set of blinking Christmas lights or a few pie tins on a string, clanging together in the wind, is another cheap and easy way to shoo nasty beasties. As a rule, most would-be raiders don't like the idea of too much noise and activity at the scene of the crime. A bashing of tin and a light show may be more effective than the occasional rifle threat.

Roosters are great protectors of hens. See page 92 for the pros and cons of having a rooster in your flock.

It was Chicken Day. I was up at six o'clock and anxiously awaiting a phone call from my chicken mentor, Diana, who runs a small organic farm about 30 minutes away from my home. The call soon came: 105 chicks had been delivered to my rural post office, waiting for pickup and their new homes.

Making my way through snowy roads, I pulled into the unplowed post office parking lot, retrieved the crate of chickens, and loaded them into my station wagon. They were *so loud!* You just can't know! Reaching Diana's house, I pulled around back as instructed and honked. She came out, all smiles, and we hauled the crate inside.

The basement of Diana's homestead is really more of a workshop. Seedlings under lights, tools and feed sacks, a prep room for slaughter and plucking, and all of it monitored by a fat gray cat named Agatha and a black setter cross named Angus. She set the crate on a workbench in the furnace room where her chicks would reside. I was ridiculously excited. I'd never watched anyone open a box of 105 baby *anythings*.

The lid came off and erupted in a cacophony of new-chick noise. I scanned for my five black chicks in the sea of yellow roasters and brown layers. My chickens were a Japanese breed called Silkie; they all have naturally black skin, bones, and muscle. I wanted ones with black feathers, too. Silkies are categorized as "bantams," which means one-third smaller than a standard-size chicken. Finally I spotted them: the little Emilio Gonzalezes of the chick world.

After hatching, a chick can live for up to three days off the nutrients of its birth egg. Now, having survived two days without food or water, they were stressed, tired, and thirsty. Chick by chick, Diana and I dipped each beak into one of the chick fountains inside their wire and blanket home. Once the birds had been shown the way, they would settle in by their canteens and drink whatever they needed. With the birds all set to go, we heated up a water bottle, covered my babies up with a handkerchief, and I drove back to my farm.

It was the beginning of a new life for all of us.

SAFE AND HEALTHY CHICKEN HOUSING must be easy to clean and requires the following:

- Adequate space
- Proper temperature
- Good ventilation (but no drafts)
- Sun/wind protection (but plenty of light in the daytime)
- Barriers against rodents, other birds, and predators
- Adequate roosting space (see Roosting 101 on page 65)
- Clean nests (see Nesting Box 101 on page 95)
- Sanitary feed and water stations
- Good drainage

Minimum Space

Chicks and chickens have different needs at different ages and under different conditions. Here are some general things to know:

- Chicks younger than 1 week should not be outside.
- As chickens grow they need more space (see chart below).
- Chickens who live outdoors during the day need less indoor space per bird when they do come in to roost.
- Chickens who are always confined need more space per bird.
- In regions with strong winds, keep the coop less than 4 feet high and stake it well.

Proper Temperature

Until your chicks have feathers (at about 20 days old) they need a reliable source of external warmth, such as a heat lamp. Keep a thermometer at chick level in the brooder and for the first week maintain the temperature at approximately 90°F (32°C). Each week reduce the heat (by raising the lamp) by approximately 5°F (2.5–3°C) until it reaches ambient temperature.

For adult chickens, the ideal temperature is 70–75°F (21–24°C). They will not thrive (and may suffer) in a temperature above 95°F (35°C).

Space Requirements for Chicken Housing

This chart shows space needed by free-range or penned birds. In an urban or suburban backyard, however, you may need to keep your chickens in a cage. A height of 24″ is adequate; add another 6″ for roosts. For one bird, a cage should be at least 30″ wide by 24″ deep; for two birds, 27″ by 32″; for four birds, 46″ by 32″.

HEAVY BREEDS

AGE	SQ FT/ BIRD	BIRDS/ SQ M	SQ FT/ BIRD	BIRDS/ SQ M
	FREE-RANGE		PENNED	
1 day–1 week	—	—	0.5	20
1–8 weeks	1.0	10	2.5	4
9–15 weeks	2.0	5	5.0	2
15–20 weeks	3.0	4	7.5	1.5
21+ weeks	4.0	3	10.0	1

Ventilation

Chickens can easily succumb to respiratory ailments without adequate ventilation. To release warm air, provide small holes near the ceiling of the coop. Screens or hardware cloth will keep out wild birds, and a cover on the north side will block the winter wind.

Flooring

Several options exist for flooring, including:

Floorless. PROS: Least expensive; no need to change bedding; birds can peck at natural ground without fear of predators. CONS: You must move the coop every couple of days so that the manure doesn't build up. TIP: Make sure there are no dips in the ground that could admit predators or a flow of water.

Solid floor (wood or concrete). PROS: Sturdiest option; offers best protection from predators. CONS: You must add bedding and change it regularly. TIP: If it's a mobile coop, move it once a week to preserve the ground underneath.

Wire floor. PROS: Excellent drainage; protects against predators. CONS: You must clean the wire regularly and still move the coop once a week. TIP: Not suitable for heavy breeds because the wire harms their feet.

Bedding (a.k.a. Litter)

Chicks need bedding at least 4 inches deep; chickens, twice that.

- Top choice of bedding is usually pine or other wood shavings. They are light, dry, absorbent, fragrant, the right size — but can be costly.
- Excellent alternatives are chopped straw and well-dried, chemical-free lawn clippings.
- Shredded paper (newsprint) is another good choice but must be replaced more frequently.
- Hungry, curious chicks may start to eat wood shavings. Cover bedding with a layer of paper toweling for the first week to prevent this.

Recycling and Retrofitting

You can make a stationary coop out of a small wooden building such as a child's playhouse or toolshed. And a camper shell can turn into an excellent portable structure.

LIGHT BREEDS

AGE	SQ FT/ BIRD	BIRDS/ SQ M	SQ FT/ BIRD	BIRDS/ SQ M
	FREE-RANGE		PENNED	
1 day–1 week	—	—	0.5	20
1–11 weeks	1.0	10	2.5	4
12–20 weeks	2.0	5	5.0	2
21+ weeks	3.0	3	7.5	1.5

BANTAM BREEDS

AGE	SQ FT/ BIRD	BIRDS/ SQ M	SQ FT/ BIRD	BIRDS/ SQ M
	FREE-RANGE		PENNED	
1 day–1 week	—	—	0.3	30
1–11 weeks	0.6	15	1.5	7
12–20 weeks	1.5	7	3.5	3
21+ weeks	2.0	5	5.0	2

Information adapted from *Storey's Guide to Raising Chickens* by Gail Damerow (Storey Publishing, 2010).

chapter 4

Poultry Prep

Preparing the Flock's First Home

Regardless of where your chicks are from — and whether you're raising 4 or 40 — you'll need to prepare for your new tenants way before the farm, feed store, or post office calls to say come and get 'em. The first weeks of life build the foundation for your stock's overall health, and a few commonsense and minimal preparations will be crucial to the birds' development and well-being. I mark my calendar three days ahead of an anticipated delivery date as "Chick Prep Time": three days before the post office or feed store calls, I'll have all the basics on my supply list. These include:

BROODER

Your little girls just emerged from a very warm previous address — the underside of a mother hen — so you'll need to mimic this as best you can. Enter your chicks' first home: the brooder. To be a proper brooder, 90°F (32°C) is the magic temperature, and you'll want to have the heat source pumping at least a day in advance to make sure it warms up to that level — and maintains that warmth.

You'll need a clean, comfortable, and warm place, free of drafts and dampness — the quickest way to end a chick's life is by putting it in cold and damp quarters. I always use the bathroom as my default brooder room. Most bathrooms have plenty of sockets for a heat lamp; floors conducive to scrubbing; a sealable door to guard against curious cats, dogs, ferrets, and four-year-olds; and it's a place you're certain to visit several times a day — convenient for making observations of temperature, food and water supply, and overall goings-on.

The brooder can be as simple as a cardboard box or as elaborate as an old claw-foot bathtub in a corner of the garage (providing your garage isn't damp and drafty). Just make sure the sides are at least 18 inches high so the birds can't escape as they grow. A chick that manages to

fly out might not be able to get back to her food, water, and — most importantly — heat source. What might seem somewhat comical on a warm July night would be a calamity on a cold April morning. To make sure your chicks are securely confined, consider fastening a screen over the top. This added safety measure will also deter too-curious house cats.

ELECTRICITY

Unless you live in the jungle with a daily temperature that's already 90°F (32°C) or warmer, you'll need a dependable source of constant electricity. When I say dependable, I mean it. So make sure you have your nursery planned for a location with a good electrical circuit and in a spot you can check on before you head out the door for work in the morning. While your electrified toolshed might at first glance seem like the perfect location for the brooder, if the power to the shed tends to flicker and go off, one cold night could kill your chicks.

HEAT BULB

The heat bulb turns an ordinary box o' chicks into an incubator. Though for years people have used 100-watt bulbs to raise chicks at home, I'd strongly advise an ultraviolet heat bulb that's specifically manufactured for use with animals. Costing only a few dollars more than standard bulbs, ultraviolet bulbs (which come in red or white) do a far better job of keeping animals warm, and the heat source is consistent for the life of the bulb, which may last you a few years!

LAMP AND LAMP CAGE

A metal lamp and cage are just as crucial as the heat bulb, being integral to keeping your chicks safe. Clamp your lamp securely. For added safety, hang it from a chain or rope so if the clamp fails the lamp won't fall into the brooder and cause a fire. There must be a metal cage or wires on the lamp to keep the bulb from touching anything directly, and the socket must be ceramic, not plastic. Most feed stores carry these lamps in late winter and early spring, when hatcheries are shipping all over America. If you don't have a feed store nearby, many online chicken sites will ship them overnight.

BEDDING

Pine shavings are my problem-free favorite choice. (Cedar shavings can clog chicks' lungs, so I avoid them.) You can also use strips of newspaper or scraps of construction paper if you have that lying around. I would avoid using straw or mulch hay. It's awfully coarse and rough when you're just a couple of days old and weigh only a few ounces. Imagine placing two-day-old humans on a pile of bamboo rods. Same thing.

You can clamp your heat lamp to a board laid across the top of the brooder box.

CHICK WATER FOUNTS

Chick founts are small and inexpensive and worth the investment. Don't try to save a few dollars by buying adult chicken founts they'll "grow into." If one of those gallon containers spilled, your birds could die from the dampness or possibly even drown. You can get small plastic two-part founts or buy the ones designed to attach to mason jars (which I prefer). They're made of sturdier metal and can utilize old pint or quart jars you have stored. Just make sure to sanitize the jars first by boiling them before filling them with your chicks' water.

CHICK FEEDERS

Like the founts, you should opt for baby-size feeders. They come as small plastic tumblers, screw-on attachments for mason jars, or trough-type feeders, which I'm actually quite fond of. The ones meant for chicks have little holes that force the birds to keep some sort of inline order during a feeding frenzy. Yes, you can always keep it simple and just place a pie tin in there, but chicks are messy eaters. Brooder cleaning is a lot easier when you have a specific and orderly place for chick starter feed.

Feeders and Founts

THOUGH YOU MIGHT BE TEMPTED to improvise when it comes to feeders and waterers — using such things as small dishes for feed or shallow plastic containers for water, you'll raise happier, healthier birds if you purchase regulation chick feeders and founts. Open feeders are too easy to soil (read: poop in), and open water dishes are too easy to stand in (or even fall into), risking a fatal chill.

CHICK STARTER FEED

Besides the correct type of brooder, this is the most important item on your chick list. You can buy medicated starter feed or organic. Be warned, though, that if you opt out of the medicated feed, you'll need to prepare for the most common and significant ailment affecting chicks: coccidiosis (see box opposite and description on page 121). Good management and early treatment are the best ways to avoid this heartbreaking disease.

To help your birds grow up healthy — or to grow up at all — I advise going with medicated feed from day one since the medicated feed is a coccidia preventive. While there is some value in a completely organically raised chick, it may be more for *you* than your chickens. If you want organic eggs, you can always switch to a regime of organic feed when the girls are laying. Your chicks will thank you.

FYI, starter feed is higher in protein and lower in calories than adult (a.k.a. "layer") ration; conversely, layer ration is higher in calcium. This higher percentage of calcium can seriously damage a chick's kidneys, so adult feed should never be substituted for chick feed — even in an emergency.

Medicated chick starter will keep your babies free of lethal diseases.

Developer feed is for pullets, your adolescent chickens.

These pellets are for layers. Layer feed also comes in mash or crumble form.

There is specially formulated feed for each cycle of a chicken's life. Never substitute one for the other.

CHICK GRIT

Birds ingest grit to aid in their food digestion, and chicks are no exception. If your chicks were out in a barn with their mother hen, they'd follow her around, watch her pick up a buffet of items, and mimic her actions. A blade of grass here, an ant there, some scratch grains to nosh on, and then a few tiny stones to store in their gizzards to help pulverize their food. But since they're being raised in a brooder rather than a natural environment, you need to supply them with that odd roughage yourself. We'll talk more about this later, but know that grit will be a good thing to have on hand in a few days when the birds are kicking dust and pine shavings around, looking for those rocks. You can buy grit in small bags or by the pound at most feed stores.

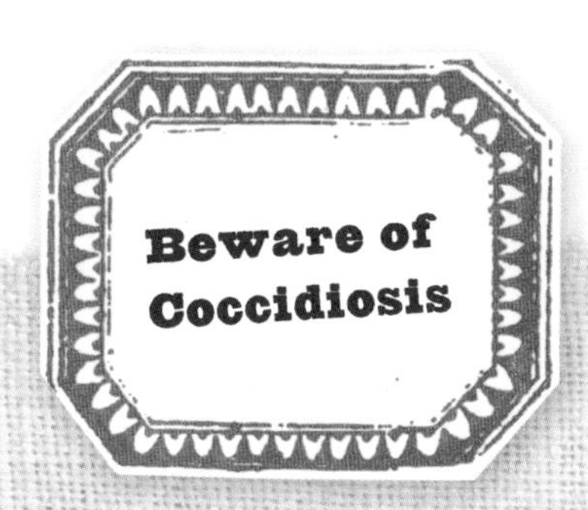

Beware of Coccidiosis

COCCIDIOSIS IS A VERY COMMON INTESTINAL DISEASE caused by the parasite coccidia that can be lethal to your vulnerable babies. Chicks pick up the protozoa by eating droppings in the feed or water or in their litter. If you notice blood in the birds' droppings, they have coccidiosis. Quick action is required!

You need to promptly clean out the entire brooder and replace the bedding with fresh stuff (maybe replace the cardboard box itself if that was your brooder route). Chicks that have coccidiosis must be treated immediately. Have on hand an anticoccidial (or coccidiostat), available online or at your feed store, for prompt treatment. If started at the first signs of the disease it is effective.

As a precaution you can add 1 tablespoon of apple cider vinegar per quart of drinking water, from day one — especially if you prefer to feed organic food from the get-go. Routinely, you should also thoroughly wash the waterers every time you refill them.

THE DAY BEFORE

Set up your brooder the night before your chicks arrive. Use a thermometer to make sure the temperature stayed steady overnight. Fill and spill a water fount to make sure it won't create any standing water, and be sure you have a good supply of chick feed and chick grit on hand.

The first day with your chicks will be a hectic blessing but certainly not an unmanageable one. There will be a flurry of activity as you welcome the kids into their new home, but they'll be a calm pile of fluff before you know it. If you did your homework — and your preparations — you're golden.

When I stopped at the feed store after work today I heard that wonderful sound of day-old chicks chirping in the back room. I bolted to the front desk to ask if there were any extras for sale (the hatchery sends a few extras in case any get lost in the mail). They had only two: a Rhode Island Red pullet and a Black Silkie bantam.

Now, I hadn't been around Silkie chicks since Idaho, and instantly those feelings of getting my first-ever laying hens flushed into me like an injection of warm nostalgia. I missed Diana and felt like I was once again in her basement during a March snowstorm getting my first order of birds. And good lord, I'd forgotten how small they were. I paid the nice people $3.60 for the little babe and took it home (along with the orphan Red).

chapter 5

Welcome Home, Girls

Week 1

MEET THE TRIO

It's time that you're introduced to the girls: Honey (a Buff Orpington), Tilda (a Rhode Island Red), and Amelia (a Blue-Laced Ameraucana). Throughout the following chapters, we'll follow these same birds from hatchlings to laying hens, and it just so happens that they represent a perfect variety pack for a starter course in poultry raising. We have a rep from the heavy breeds, another from the die-hard reliables, and a prime example of the illustrious exotics.

HONEY

Buff Orpington
Sassy Yet Docile

ORPINGTONS LIKE HONEY ARE A LARGE, ROUND TYPE OF HEN suited for long, cold winters. These sassy, strong, bug-eating birds originated in the village of Orpington in Kent, England, where they were bred to handle harsh weather, lay large brown eggs, and assist the small producer in heavy meat production. Being a dual-purpose fowl (suited for both meat and egg production), Orpingtons give the homesteader a lot of bang for the buck.

Though Orps are one of the biggest birds in the laying world, their docile nature makes them good with children, calm with grandparents, and patient with cats. These big softies (literally — these birds have some of the softest, most beautiful golden plumage I have ever had the luck to feel) will let you scoop them up and will eat out of the palm of your hand. There's a reason so many folks love the Orp, and Honey is no exception.

Honey at two weeks old

Five weeks old

Seven months old

TILDA

Rhode Island Red

An American Classic

TILDA'S THE HARD-WORKING FARMHAND of the bunch and rightly so because these girls come from farm-bred stock. Unlike other breeds raised for looks, these gals were bred to produce, produce, produce. Developed back in the 1800s for both egg and meat production, these New England natives may not look as fancy as the other gals, but they can lay like nobody's business. I like to think of Rhode Island Reds as the Rosie the Riveters of the chicken world. Anyone who's had a few of these industrious and independent birds would agree. She's also the only breed of chicken to receive the honor of being a state bird! (I'll let you guess which state.)

Tilda at two weeks old

Five weeks old

Seven months old

AMELIA

Ameraucana
The Easter Egg Hen

AND THEN THERE ARE GALS LIKE AMELIA, the Ameraucana. Of South American heritage from stock imported in the 1930s, this breed was developed mostly in North America in the 1970s. Though new to the scene, Ameraucanas are rapidly rising in popularity because these birds have quiet souls and quirky features. They coo like doves and walk around your yard with slate or blue-green feet and puffed-out feathered cheeks. They come in many gorgeous varieties of plumage, and as a bonus, they mix up your egg basket's spectrum since they lay eggs in a variety of hues from light blue to pink to green. This grand trick has earned them the nickname of the Easter Egg Hen. I've had my share of Ameraucanas and always appreciate those pale jewels in the nest box among the creams and browns of the other girls' eggs.

Amelia at two weeks old

Five weeks old

Seven months old

Acclimation

As soon as you get them home, you'll need to get the chicks acclimated in three simple steps:

1. One at a time, pick up each chick and gently dip her beak into the waterer. Watch for her to swallow. A chick shouldn't go more than 48 hours without water after hatching. The sooner they drink, the less stressed they'll be, and the better they'll grow. It's important that the birds drink before eating since this seems to help prevent sticky bottom (more on that later).

If the chick takes the hint, she'll smack her mouth and throw her head back to drink.

2. When she's satiated with water, it's time for food. Place her in front of the starter feed, and watch as she discovers that food is exactly what she has craved over the past 48 hours since leaving the egg.

3. When every member of your new coop has been given a long drink and eaten a good meal, all that's left is to assure that their brooder is warm and dry and located out of harm's way.

Don't be surprised if the chicks fall asleep right away. Like human newborns, chicks sleep often but never when you want them to.

Chick Diary

We started with a recycled egg carton for the chicks' food and a small plastic dish for their water, but we immediately recognized the need to get a regulation chick feeder and fount. Baby chicks don't have a lick of common sense and will stand in the water, get themselves wet, and catch a chill. Plus they poop on everything in their environment and kick litter everywhere, including their food and water, given the chance.

Containers specifically designed for chicks help keep their food and water supply clean — but I find that I *still* need to excavate their food from piles of wood shavings at least twice a day.

Tilda falls asleep face first after arriving home.

Despite dire warnings from all the chick lit we read of kicked-up dust from chicks' scratching and their loud, near-constant peeping, we decided to start our newborn chicks in the house. We wanted to spend some serious time with our new pets in the comfort of our home rather than in the unfriendly space of either the garage or the basement.

We set up a cardboard box lined with wood shavings in our guest bedroom, and it worked out great — probably because we had just three birds. (After a month, a light layer of dust kicked up from the wood shavings coated the windowsills and other surfaces, but it was worth a little extra cleanup to comfortably hang out with the chicks during those first special weeks.)

Since our adventure included our seven-year-old son, we wrote some rules for ourselves on one side of the brooder box and rules for the chicks on the other side. It's especially important for kids to wash their hands after handling chickens of any age.

Home Is Where the Hearth Is

Heat is the matter at hand. To maintain your chicks' ideal 90°F (32°C) environment, that heat bulb hanging safely above the brooder box needs to be positioned much as a mother hen would be: ever present and warm but not a hindrance. It will be quickly evident when the biosphere is happy. If all is well, the chicks will meander under the lamp as if it were the sun itself. They'll wander from fountains to feeders in mindless, contented chickdom.

But if your bulb hangs too high, making the brooder temperature too cool, all of the gals will be huddled in a pile under the light, trying desperately to keep warm. If you notice a fluffy huddle, lower the bulb 2 inches, and check back in half an hour. Be aware, though, that the brooder can be too warm as well. If the birds are pinned against the corners and walls of the box, trying to escape the heat and each other, raise the lamp 2 inches. Keep adjusting one way or the other until the birds appear content. The goal is a blissful state of bird being. It'll take a little practice, but perfect lamp placement is key in raising your chicks to be healthy layers.

When the heat lamp is at the right height, chicks will neither huddle under the light nor hide from it.

There are moments when you realize your life has changed and you are never going back — couldn't if you wanted to. For me that realization came about because my coworker Noreen wanted some laying hens.

We found a backyard chicken keeper on Craigslist who was thinning out her flock. Since I lost five birds over the winter and Noreen needed a starter set, we struck a deal and a pick-up time, and I filled the back of my station wagon with wire cages. Come lunch break, Noreen and I were driving to North Bennington to meet our new livestock. Though it had been pouring all morning, we did not waver.

I found our destination and pulled into the driveway. Two border collies in the window of the white farmhouse barked a suspicious welcome, followed by a young woman who emerged in Wellies and a raincoat. The rain picked up as we stood in the big fenced-in run and she pointed out which birds could go and which would stay.

I knew I'd have to catch these hens fast or return to the office looking like a refugee. Noreen watched from the dry comfort of my car, laughing as I scurried around hunting and trapping our new acquisitions. Now soaking wet, I scooped up the chickens, cradling one or two at a time in my arms as I ran back to the car. They carried on something fierce, despite my reassurance that their new homes would be a long call from the Purdue factory, and they should lighten up.

As I drove back into my office parking lot, a melody of clucking coming from the back seat, I thought to myself, *This is my life now.* And I grinned. A girl from Pennsylvania who fell in love with homesteading in Tennessee pulling into her Vermont office's driveway with a carful of chickens, aided by an Idaho poultry education.... Like I said, I like the story so far.

Handle with Care

Who can resist picking up a days-old chick that's all fluff and nonsense? While they're still adorned in their baby plumage (they have yet to grow their first pinfeathers), take a moment to handle them and feel the gentle softness of their down feathers. But handling the chicks is as much about getting a sense of their spunk and health as it is of their adorableness.

A healthy bird will feel like a miniature dinosaur in a bunny suit: all hearty cuteness with ribbed taloned feet and bright beady eyes. Handling your chicks also gets them used to people and accustomed to being touched. Raising just a few birds allows you to truly imprint on your flock and establish a personal relationship. I am not suggesting carrying your Buff Orpingtons and Jersey Giants around in your pocket, but daily gentle handling will let the birds know a hand isn't a death threat, and ongoing physical contact means maturing birds will likely be less flighty.

The more time you spend handling your chickens, the tamer they will become.

Our bearded lady Amelia has always been shy and quirky. As a chick she started the odd habit of rubbing her beak from side to side like a knife on a sharpening stone.

Orpingtons are brawny but usually quite mild-mannered. Honey, on the other hand, uses her impressive bulk to assert her place in the pecking order.

Getting to Know You

As your new birds adapt to their brooder, they'll ease into a regular routine of eating, exploring, and resting. Pay close attention, and you'll notice their individual personalities emerging. You'll see which of the lot will plow headfirst into the feeder when it's filled in the morning, which chick may keep to herself or lag behind in the chase when a moth flies into the box, which chick never seems to stop peeping, and which won't make a sound unless you reach a hand in and pick her up.

And as the animals in your little biosphere get comfy with their new world, they'll also be getting comfy with each other. One of my favorite moments with a new hatch is when they're all asleep in a toasty pile, breathing quietly under the life-sustaining warmth of the heat lamp. In a few turns of season, they'll all be lined up side by side on a roost, keeping warm in practically the same way. Safety (and comfort) in numbers: Even the little ones seem to know.

From day one, Tilda was bold and sassy.

Time to Clean House

After four days of chick use, your brooder will need some basic housecleaning. Ready a small cardboard box with clean bedding and a lid, and relocate your chicks into it one at a time. Swap the brooder under the heat lamp with the temporary condo. While the residents are exploring their vacation home, take the brooder outside, remove all the old bedding, and check the container itself. Is the bottom wet or the woodchips damp? If so, there might be an issue with leaking water (which could be a dangerous health risk if your birds get damp).

If you are using a cardboard box, make sure that the bottom is still sturdy. After a while the bottom of the box can get hideous, and the last thing you want when you lift it to take it out and clean it is to discover that a gross amalgam of decaying woodchips, chick poop, and wet cardboard has made a home on the floor of your bathroom — or, even worse, your spare bedroom.

Tip: Place feeder and fount on a wooden board to raise them slightly above the bedding surface so the chicks will track fewer shavings into their food and water.

TRUE GRIT

When the brooder has been cleaned, the water founts have been washed (with mild, all-natural detergent), and the feeders have been cleaned and refilled, fill the box with fresh bedding, and sprinkle a small amount of chick grit among the shavings so they'll have something to scratch and claw for.

You haven't introduced grit yet? Well, this would be a good time. The birds are now accustomed to their surroundings, thoroughly get the whole food/water bit, and will be able to physically (and mentally) swallow those tiny stones you bought from the feed store. Grit acts like a grinder in your chick's gizzard. The coarse sand and tiny pebbles help break down the soy meal and corn mash in their gullets. If the chicks were being raised outdoors by a mother hen, the ground would provide them with all the tiny knickknacks their crops could ever want. However, brooder-raised gals don't have that option, so a small supply of these indigestible but important little rocks will help keep your flock healthy and growing.

Okay, you can return your chicks to their freshened home now.

Chicks will scratch wood shavings into the fount. Clear the trough of litter and provide fresh water at least twice a day.

The lid helps keep the food clean, but the chicks will still manage to soil it. Scoop the poop and wood shavings out of the feeder twice a day.

A Word about Hygiene

WHEN YOUR BIRDS ARE CLEARLY THRIVING in their brooder, it's easy to drop your guard a little and not worry about the small things you once obsessed over. Temperature regulation is under control. Food and water sources are up to par. And the birds now seem to be used to their unnatural surroundings and constant daylight. Well done. But don't pour any celebratory drinks just yet. Your girls aren't out of the woods for quite some time. Pullets and cockerels under two months of age are extremely vulnerable to diseases caused by lax care or hygiene negligence.

Make sure you set aside time at the beginning of each week to really clean out their starter home. Adding clean pine shavings to the existing pile may get you through the week, but this negligence will eventually take its toll. What seems clean on the surface may be hiding a mess of bacteria and germs lying in wait for your birds. If you're not careful, you'll just be mulching a disease pit because below those fresh pine shavings is a stew of damp, moldy feed and muck. Not good. You're just asking for coccidiosis to take your birds.

Easily avoid this by throwing out all those old pine shavings and laying down a precut piece of cardboard as an absorbent base in the bottom of the box. Even better, replace the original cardboard box with a new and even slightly larger one each week as the birds grow.

Peaceable Kingdom

Growing up healthy and strong means your layers can provide a lifetime of service to you as egg producers. We already covered the importance of a clean, dry environment to prevent chill or disease that a damp home can cause these fragile birds, but I also want to emphasize the value of a stress-free life for the young ones.

The brooder should not be placed where pet cats skulk, radios blare, dogs bound, or kids relentlessly scrutinize. While all of these sensations are good for your birds to grow up around in small doses — skulking cats being the exception — they should not be a constant presence. If chicks get too stressed out, they could literally die from it because their vents will seal up — and this is not a good thing.

YOU DON'T WANT SEALED-UP BUMS

What?! OK, that was a little blunt. But it's true. A stressed-out chick can get diarrhea — and swilling Pepto-Bismol isn't an option. What happens when a young chick (under fourteen weeks, especially) gets the runs is that the excrement pastes up and dries over her vent, clogging the passageway and making it impossible for the hen to relieve herself. Plain and simple, if pasting up happens, she'll die — and it's a sad way to go.

So while your chicks are still in that fluffy vulnerable stage, periodically pick them up and check their vents. Make sure there isn't any dried poop to cause serious damage. If there is, you'll need to get a warm cloth and wipe it off till the area is clean, or cut away the pasting with carefully wielded scissors.

Check your chicks regularly for pasted-up vents. It could save their lives.

There isn't a chick in the world that won't scream while you do this, but it is certainly for its own good. A watertight ending to a short life won't make anyone's day brighter. So keep an eye on all aspects of your animals' ups and downs. It could save a life!

PICKY CHICKS

You may have beautiful heritage chicks, the world's cleanest brooder, and a heat-monitored room, but that doesn't mean trouble can't arise in the most unlikely of places: your birds themselves. Sometimes chicks peck each other to the point of bleeding or (heaven forbid) cannibalism. A little pecking or the occasional bald spot on a lower-ranking bird is normal, but if blood is drawn, you need to separate the victim from the rest of the flock.

To prevent bullying, make sure your birds have plenty of space, food, and fresh water. Stress can cause violence among young birds. If they are content, they are far less likely to choose a scapegoat. Another option is to use a red brooder light, which tints the entire home red and may hide the sight of fresh blood.

Perching is to chicks what first steps are to babies — an important milestone in their development. Some chicks will have the confidence and balance to try it as early as the end of the first week.

We built a beginner's perch by nailing together three sturdy tree branches in an H shape and leaning it against the side of the box. Brave Tilda was of course the first to hop up. Amelia looked a bit seasick when we coaxed her on.

Roosting 101

ELEVATION IS NATURAL FOR CHICKENS. In the wild these birds fly up into trees to be safe from ground predators. Even your little ones will feel this instinctive draw and hook their talons around the roost. All that's needed for your chicks' first perch is a thin branch or dowel that can be situated an inch or so above the birds' bedding. This simple addition literally elevates the chickens' status. Besides giving them a perch to learn balance and roosting on, it invites a new level of play and social interaction in the brooder.

Watch your chicks learn to jump, fly onto, and launch themselves from the roost. See how certain chicks always stick together, even squat side by side on that branch. It'll be a welcomed addition and get your chicks accustomed to what will surely be their future sleeping posture of choice.

Adult birds, too, need to be put on a pedestal. Roosts are a must-have in your coop plans and should be at varied heights for the birds' preferences. My coop has windows, rafters, and near-floor-level roosts, and each bird seems to have an altitude that suits her. Some birds will go as high as their wings will take them, and others just want to be on the same rung as their rooster. Whether your coop has a few branches screwed into the wall or a dramatic system of wooden pallets in pyramids matters less than that the animals have options of height and space.

Fledglings

By the end of the first week, your chicks' baby fluff is molting quickly. They're developing wing feathers — which look like feathered car doors — and the tail feathers are beginning to poke out of their bums.

Their behavior is rapidly changing, as well. Drop a wad of tissues in the brooder and you'll notice immediate response and reaction. They'll duck, run around it, or walk up and explore it. Their minds are getting sharper, and so is their self-awareness.

As week one ends, feathers begin to replace fluff on the wings and tail.

chapter 6

Growing Up

Weeks 2–7

Growth Spurt!

Obvious physical changes occur in the second week of a chick's life. And with the physical changes come changes in their habitat.

APPEARANCE

At this stage, the girls' feathers are quickly growing in, and they're looking more like regulation chickens. Pinfeathers are visible, and wings are rapidly developing, with considerable test flapping taking place. Tail feathers develop early in the week. Neck feathers will develop by the end of the week.

TEMPERATURE

It may very well be time to tweak the heat in the brooder. In their first week of life, chicks can't regulate their own body temperature, but by their second and third weeks of age, they can. To help them in this process, raise the heat lamp 3 inches. That slight height difference will lower the temperature in the brooder a few degrees, and the chicks will learn to adjust their bodies to the somewhat chillier space. It may seem like tough love, but it's necessary. The brooder should still be warm (nothing

You can see growth not just from one day to the next but even within the same day! They can't fly yet, but they're not above clumsy attempts.

cooler than 85°F [29°C] this week), but that slight change of temperature will help pin-feathers form and bodies grow healthier.

As a general rule of thumb, the brooder temperature should be lowered by 5°F (2.8°C) with each successive week. By now you should have a pretty good handle on the temperature situation and can start to train your eye to know what's right.

RELOCATION

If your girls are still living in their original baby box, you may want to upgrade to a larger brooder. (Some folks start with a large container from the get-go, but I like letting the boxes grow with the flock, which keeps them together in a more manageable space.) Changing brooders weekly also gives them a clean, fresh new home — and as you now know, a clean brooder is a healthy brooder.

Like buds on a plant opening into flowers, shafts sprout from the chicks and new feathers unfold.

I don't have a lot of spare time — when I come home from work, the farm gets all of my attention. The sheep need hay. The goat needs fresh water (in the bucket he inevitably kicked over). The chickens need their evening scratch doled out and their eggs collected. And the dogs need to be walked and fed. By the time everyone and everything's been seen to, a high form of relaxation is in order. I mean music. I mean picking up an instrument and playing.

I don't play for anyone in particular, usually just for the evening herself. And I'm not a great musician by any means, but I get by. I play some guitar, a little clawhammer banjo, and a beautiful mountain dulcimer engraved with leaping deer. But when the day gets really long, nothing soothes me like the drone of my fiddle.

When I started playing old-time mountain music and researching the history of the ballads and tunes of the South, I came across a lot of odes to chickens. Horses and pickups may be the stars of modern country music, but chickens were the beaus of the old-timers' hearts. Fiddle songs like "Chicken Reel" and "Cluck Old Hen" were the hits of their time — and the musical versions of the birds themselves. They're tense and quirky reels that make you want to strut like a rooster. Playing them always brings a smile to peoples' faces, even the unfortunate ones without a henhouse to call their own.

Chicken Reel

Cluck Old Hen

My old hen's a good old hen
She lays eggs for the railroad men
Sometimes eight and sometimes ten
That's enough for the railroad men

Cluck old hen cluck and squall
Y'ain't laid an egg since way last fall
Cluck old hen cluck and sing
Y'ain't laid an egg since way last spring

Cluck old hen cluck when I tell you
Cluck old hen or I'm gonna sell you
Last time she cackled cackled in the lot
Next time she cackles cackle in the pot

My old hen she's a good old hen
She lays eggs for the railroad men
Sometimes one sometimes two
Sometimes enough for the whole damn crew

Whenever I cup Tilda in my hand and sing to her, she falls asleep. It works every time. Hens sing when they are happy and to amuse themselves, just as we do. I think when we sing to them, they pick up our cues for relaxation and contentment . . . depending on the playlist in your head that day, of course!

Chick Diary

As the chicks lose their baby fluff to brand-new feathers, bald patches appear, especially under the wings and belly.

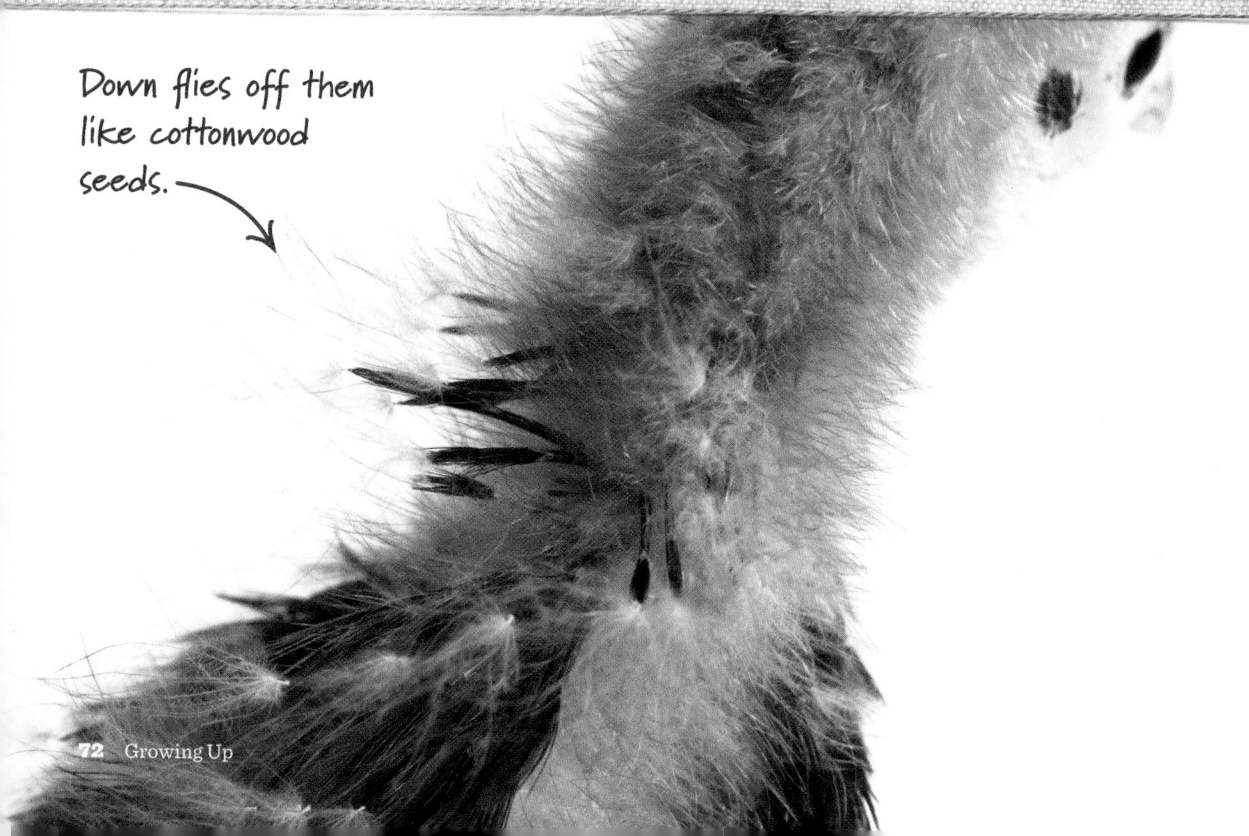

Down flies off them like cottonwood seeds.

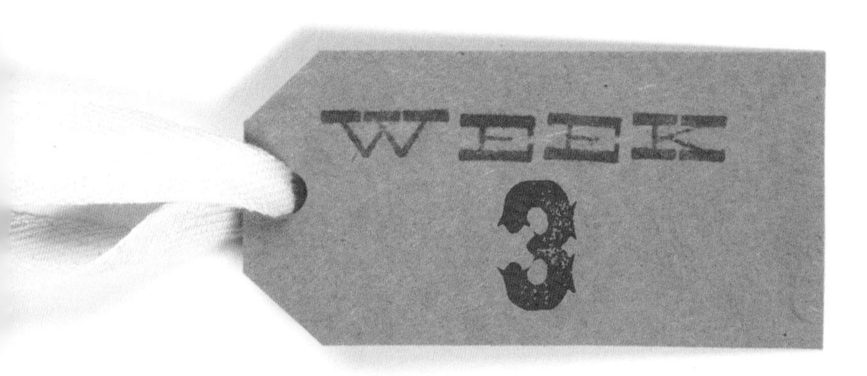

Transformers

By the third week your birds are truly starting to resemble chickens instead of chicks. Their wings, capes, and crowns of their heads have grown actual feathers, and everywhere else new feathers are either burgeoning or they're already coated in a slick new do. Sigh and smile; your girls' baby days are behind them.

You may also start to discover that some of the flock may be future poultry Olympians. With most of their wing feathers now installed, your bolder charges may take flight within their brooder or launch themselves from the tops of founts and feeders out of the safety of the nest. I usually remedy this with a fitted wire screen over the top of the box.

Combs and wattles begin to develop.

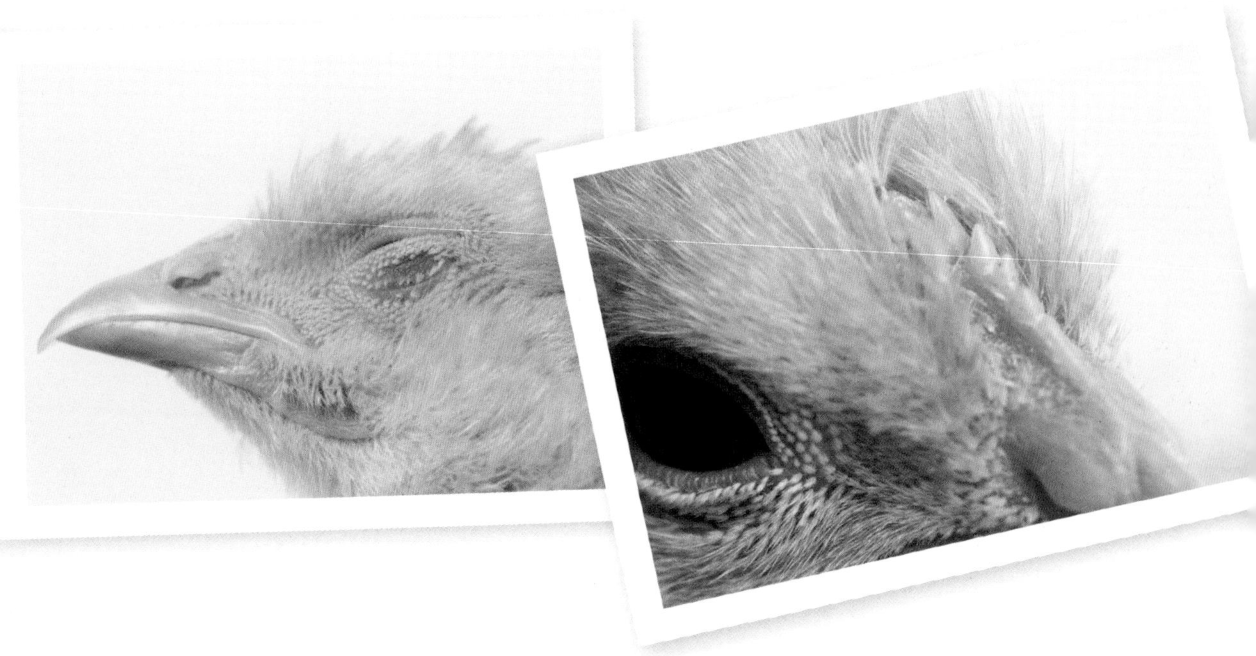

ANATOMY OF A CHICKEN

Common Comb Varieties

Single comb

Rose comb

Pea comb

The shanks of each of the girls are a different color. Honey's are pinkish, Tilda's are yellow, and Amelia's are a greenish gray.

Dust Bowl

By their fourth week of life, your chicks will start to look like miniature versions of their future selves. Most of the feathers are now in or soon will be. And they're now skilled in the art of jumping and flapping their new wings at hummingbird speeds as if determined to get out into the big world that awaits. They'll more and more gravitate toward their perches, even resting on them for hours at a spell, their tiny dinosaur feet holding on tight, learning to balance. However, these dramatic changes come with a price — and friends, that price is dust.

Oh, the dust! A thin, creamy-white film of dust will cover everything within a few feet (sometimes farther) of the brooder. It's neither toxic nor smelly, but it isn't the kind of thing you want coating your toothbrush in the bathroom. To keep your belongings clean, wipe down your surfaces with a damp rag or some natural home cleaners. You don't want to throw down industrial-strength bleaches near those tiny beaks. And make sure to periodically wipe the dust film off your heat lamp. It can clog the bulb socket or coat the wires.

Their feet seem to be as large as their entire torso was a few weeks ago.

While you're checking the heat lamp, consider it the ideal time to do your weekly assessment of the temperature situation. By now the brooder temperature should be about 75°F (24°C). Adjust the position of the lamp — likely a bit farther from the chicks — as necessary.

As Rocky's coach said, if you can catch a chicken, you can catch greased lightning. Chickens are skillful at dodging capture, and quiet, calm movements work best. We've found it works like magic to herd them slowly with our arms outstretched.

Catching Greased Lightning

SOON YOUR CHICKS WILL BE MORE THAN A HANDFUL. Already they aren't babies anymore so much as restless and rowdy teenagers, but when they are full grown, catching one and holding on to her will be a trick you'll have to master.

When you pick a chicken up and hold her, it's important to secure both wings. Otherwise, she'll flap vigorously and wrestle herself free. The more often you handle your chickens, the more likely they'll be to stand still and allow it. To avoid panic, cradle the bird in the crook of your arm, pressing one wing against your body while holding the other wing securely against the chicken. Some folks prefer to support the chicken's feet (and some just like to feel the talons around their fingers), but this isn't necessary.

Why hold a chicken? Because sometimes you need to transport them, apply medication, or just check their general health and well-being. You also want a flock that's comfortable with humans. Regular handling means your birds will be less edgy around people relaxing outside or a friend's kid who simply must hold one of the cluck-clucks.

A good tip: If you need to move your chickens to a new home, pull them off the roost as they rest at night. Darkness is the easiest time to catch and pick up a chicken, and if your coop has a light (for your benefit), you'll be home free.

Staging a Coop

In their fifth week of life, your chicks should be all feathered out in the lovely plumage of their adult selves. They're still tiny compared to what they'll become, but the markings and feathers are all there. Combs are red and healthy on foreheads, feet are golden yellow (slate or blue-green if they're Ameraucanas), and eyes are bright and constantly aware. You now have in your possession the chicken version of tweens.

Their dietary needs are changing too. If you haven't already switched from chick feed to developer feed, now is the time to do it.

A PEEK AT THE WORLD

If you live in a warm climate, consider introducing your spunky adolescents to the outdoors, and let them experience the world outside their brooder. While under your supervision, of course, and in a safe enclosure, let the birds feel grass under their feet and learn to chase ants. These girls will spend their whole lives in your backyard, so give them a taste of the sun, dirt, and delicious bugs.

I place my gang out in a big rabbit hutch in a sunny part of the yard. The wire floor is wide enough to let them eat grass and learn what dirt tastes like yet keeps them safe from neighborhood cats and dogs or the occasional fly-by hawk. (If your budget allows, the space-age-looking Eglu is an ideal habitat, being coop and enclosed run all in one.)

CHIRPING SEEDLINGS

If you're a gardener, think of this time in the sun as the chick version of hardening off seedlings. Just as you can't take a delicate tomato seedling growing under a heat lamp in the basement and instantly plant it in your spring garden without its suffering irreparable shock, chicks need to be eased into the real world. Going from that 70-degree brooder to a 39-degree night in the coop (from 21°C to 4°C) could be more than they can endure. So make these initial field trips simple and short — an hour or two in the sun in a safe place.

Use your observations of the chicks during their outdoor recesses to regulate the brooder temperature when they're back indoors. Are they having a blast outside, or are they huddled together shaking in a corner? If it's too soon or too cold, you'll know by their behavior. Trust your gut — and their body language.

Don't be surprised if your birds get frightened and fly onto your shoulder for protection. They know you better than they do the outdoors right now!

Chick Diary

Birds of a feather truly do flock together. The chickens like to sleep in a huddle, while outdoors they move through the yard in a tight group, probably from the instinct that there is safety in numbers.

Tilda became quite distressed when she got separated from the flock. Here, she shouts out to them.

Amelia has developed a pronounced beak abnormality. It is curved in such a way that she can't close it fully. (She eventually grew out of this.)

Tilda is slow to develop, even though Rhode Island Reds are usually quick to mature. Her feathers come in later than the others, and she is a runt in size. (We worried about her for nothing; she eventually caught up and was the first to lay!)

Even if developmental abnormalities had persisted, we would have loved and kept our birds. We're raising these girls as pets, so show-quality looks and farm-grade productivity are not of primary importance.

A Balanced Diet

CHICKENS, LIKE US, ARE OMNIVORES. They'll eat a finely chopped salad of your garden's greens or just as happily munch into a rare steak. When they are allowed to roam free, this behavior will take over naturally. Watch them strut around your property, diving after bugs and worms and snipping blades of grass with their beaks. Between their clover salads and insect entrees, you'll better understand the chickens' diet.

They're omnivores and should be treated as such. A pure vegetarian diet (as some egg cartons boast) is ridiculous. A chicken's gullet requires protein, and it's in the chicken's very nature to find some on its own. It's quite the sight to watch these natural hunter-gatherers at their best. I've watched my birds catch frogs, small mice, and dragonflies out of midair! However, it's not always so bucolic. They'll eat their own eggs (or each other!) given stressful circumstances. The best way to avoid egg eating and cannibalism is a balanced diet and ample space. These two things keep your stock happy, healthy, and civil.

Grain and Scraps

Most folks who keep laying hens feed a grain ration bought from the feed store. This is (usually) a soy-based protein feed that comes in many forms: pellets (like rabbit food), crumbles (pellets hit with a hammer), or mash (loose grain, finely milled). I feed my birds pelleted layer ration but try to keep the protein higher for my production animals. I want at least 17 percent of their diets to be protein and make sure the tag on the 50-pound bag supports that.

Crushed oyster shell, fed to chickens as a dietary supplement, will strengthen the shells of their eggs.

The *only* supplement you can give chicks under eight weeks of age is grit, which helps them digest their feed.

Scratch is a treat or supplement but should not replace protein-rich layer ration as your flock's primary feed.

Feed from a bag isn't all your birds should eat, however. Supplement it with kitchen scraps, and they will adore you. I keep a kitchen-compost pail by my sink, and everything from bacon bits to carrot shavings falls into there for the chickens. As long as you keep the garlic and onions away from the birds (which won't hurt them but will flavor your eggs!) you'll be turning trash into eggs. Not a bad turnaround.

Calcium and Scratch

Besides feed ration and scraps, I also provide my flock with a pie plate of oyster shell. This adds calcium to the diet and makes the shells of their eggs tough and thick. The birds appreciate the bit of vitamin-rich grit and will peck at it whenever they are jonesing for some. You just don't want to give oyster shells to chicks or young pullets; the calcium can be too much for their little systems, so save it for the big gals.

Scratch, which is a mixture of different grains — usually cracked corn, wheat, and oats — is like candy to chickens. If you want them to come to you when you call, train them by calling "chick, chick, chick" while tossing a handful on the ground. Soon enough, they'll associate your call with tasty treats. Since scratch doesn't contain any protein, which is so essential to a balanced diet, it's best used in moderation. You don't want your girls to fill up on dessert and leave no room for dinner.

The exception might be during cold winters. Think of it as stoking the fire, adding a little extra fat and energy into your birds while the temperatures drop. Your flock will appreciate some comfort food, and it will actually, physically, heat up their bodies. It never hurts to pile on the insulation come winter, and my chickens can't stand wool sweaters. So we opt for scratch grains.

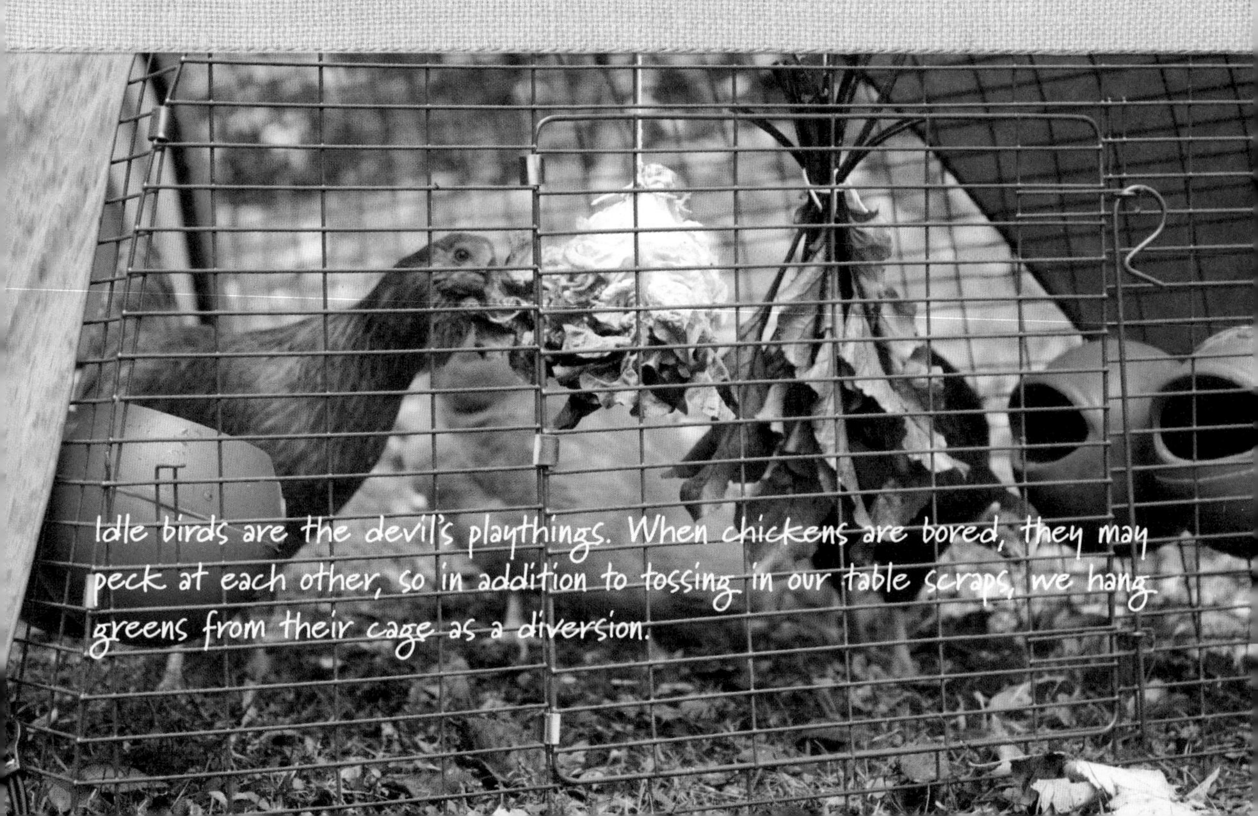

Idle birds are the devil's playthings. When chickens are bored, they may peck at each other, so in addition to tossing in our table scraps, we hang greens from their cage as a diversion.

New Digs

At six weeks of age, they're still technically pullets — and will be until they're a year old — but these kids are chickens. Lean back, take a long breath, and smile. You've raised these birds from two-day-old balls of fluff into the strutting hens they've become. And now it's time to introduce the girls to their coop.

MOVING DAY

If you live in a warm climate, or the June sun is beating down on your New England grass, moving the girls permanently outside can be a once-and-done event. Carry them out to their new home, and let them spend their first night in the great outdoors — so long as the night will be warm (at least in the 50s F [10s C]). Assuming their coop and pen will withstand predation, wind, and weather, you're finally in the home stretch. Obviously, if it's still subfreezing and snowing out, hold off until mild weather, but you'll likely be raising these girls in the spring, and a night in the 50s will be fine. Your birds are ready to meet the demands of cooler nights, roost life, and natural food scavenging, and they'll quickly learn to love their new digs.

If you're hesitant to make the big move, it's justified. You don't want one cold night to be the end of the lovely chickens you call by name. So continue to harden them off in steps. When nights are chilly, leave them in their pen all day, and bring them indoors to a temperate brooder to sleep. Replace the old 250-watt heat lamp with a 75-watt bulb, and let them learn about cooler nights in stages.

When a warm night comes around the bend, put them into the coop to roost. Check on them with a flashlight if you must, but chances are they'll be happily sleeping side by side. Having them reach independence has been the focus of all your efforts to this point.

Your girls will get to work eating grass and insects right away.

Moved the trio outside to the coop! They love the extra space and racing from one end of the run to the other.

Don't try this at home! While a lot of dogs would love to sink their teeth into your chickens' necks, Chico gets along famously with the girls.

Chico either grazes alongside the chickens . . .

or cleans up behind them!

Honey pauses to pick a piece of dirt out of Chico's fur.

Introducing Other Pets

Most people who are inclined to share their lives with livestock probably already share their bed with a cat or a dog. If you plan on letting your flock and bichon frise cross paths, you'll want to introduce them slowly. The combination of pets and chickens should be carefully considered since they can either be your own peaceable kingdom or a recipe for disaster.

The earlier you can bring your dog or cat to the chickens, the better. Showing them the brooder while keeping them secured on a leash or in your arms lets you gauge their response to the birds. Most dogs will eye the box curiously (as will most cats), but if your dogs are anything like mine, they'll lunge at the brooder, hoping for a mouthful of chicken nugget.

Be realistic about your situation. I know my chickens will never be safe around my Siberians, and perhaps your cat will never be trustworthy around your fluffball chicks. But plenty of supervision and the occasional harsh tone when a claw or jaw gets too close could be all you need to let your pets understand their new flatmates are not food.

Summertime. Enjoy the dog days with your stock. It's the time to revel in the fun that is their innate chickenness and to train them to live with you. The birds will learn to come when called for table scraps and where to look for water and shade. Grab a cold beer and a lawn chair and spend a sunset watching your birds at their most active time of day — that scurrying, clucky time before they bed down. Watch them run around and play and then, almost as if some bell went off in their brains, make a beeline for their roost and call it a night. The long stretch between now and that first egg will be a few months, so you might as well enjoy your chickens for their entertainment value (better than TV any day!).

Teen Spirit

Pullets just this side of two months old are now living outdoors, spending their days exploring your backyard and their nights safely asleep on their roost. They still look more like tiny dinosaurs with feathers than fat, happy hens, but their rumps and breasts are slowly filling out. Eventually they'll mature into the laying hens you envisioned when you started this adventure, but it'll be a few more months before they start dropping eggs.

SLUG FEST

After trying everything to protect my romaine lettuce from slugs, I discovered that my five Black Silkie bantams are the best pest patrol of all. Twice a weekend I deposit them in the fenced garden, fetch a book and a blanket, and read while they hunt for slugs. Every twenty minutes I move them to the next bed. After an hour or so, I am well read and they are well fed. I no longer have a slug problem and I can get a great tan.

Now that's teamwork.

Honey and Tilda begin to face off for dominance.

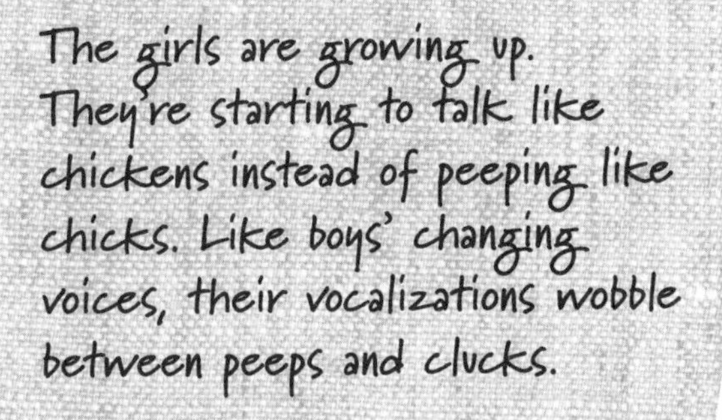

The girls are growing up. They're starting to talk like chickens instead of peeping like chicks. Like boys' changing voices, their vocalizations wobble between peeps and clucks.

They're also fighting more often, possibly to establish — yes — the pecking order.

Amelia is growing a fuzzy beard.

All the girls have sprouting combs.

Henpecked

Just like a pack of wolves or a corporate office, chickens have a pecking order. One hen will become the dominant character and make her stance in the community known — through violence if necessary. If you notice one bird constantly bossing the others around by stealing their food, being the first at the fresh water, or pushing another bird off her spot on the roost, you might have yourself a honcho. Expect any "mean girls" to emerge when your birds are about six weeks old.

While the occasional peck and lost feather patch on the back of your hens is normal, you need to be careful that your birds don't get to the bullying point of drawing blood from one of their mates. A bleeding hen can cause a craze of bloodlust and cannibalism in chickens, and it isn't pretty. Avoid this crisis by making sure your birds live in a stress-free environment. Contentment is the remedy for civil disobedience among the feathered set.

INTRODUCING NEW CHICKENS

When you bring new birds into your flock, realize that you just rocked their world. Every single bird in your coop knew exactly where it belonged on the totem pole, and by introducing a few new pullets, you just made it possible for a lot of lower-ranking chickens to jockey for a better seat.

I keep new birds in a cage inside the free-range birds' coop for at least 48 hours

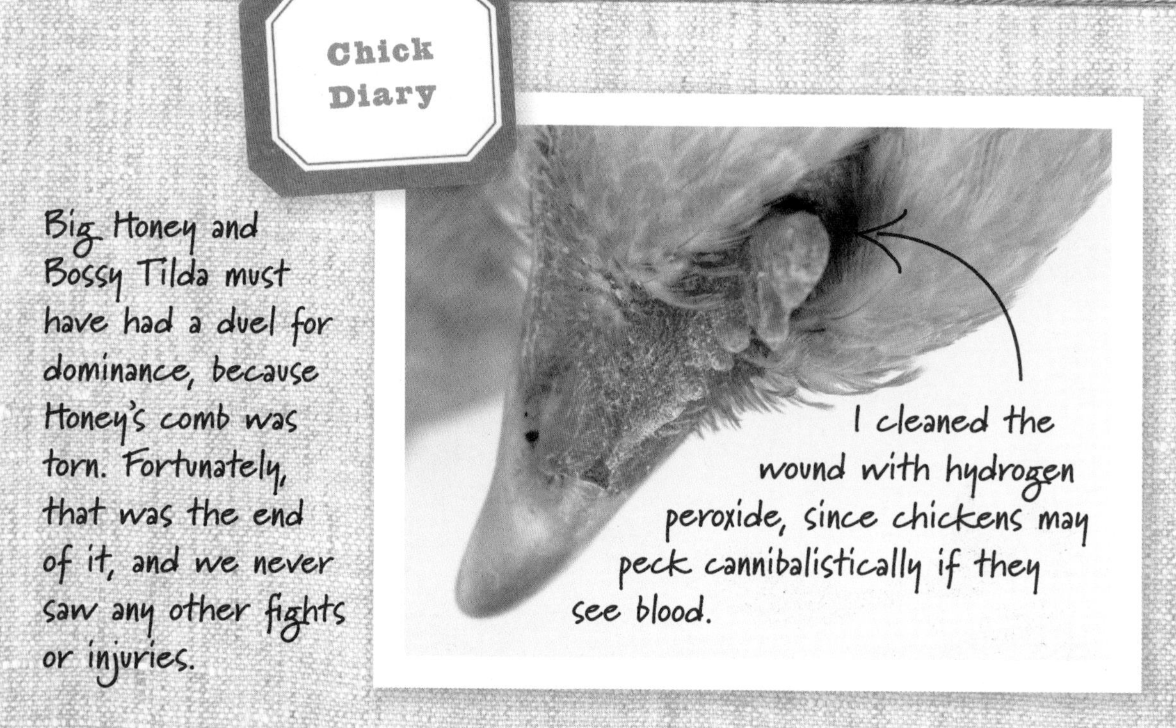

after they arrive at the farm. They have access to food and water in their cage, of course, but they also have the protection of some wire between them and any pointy beaks that may come a-knocking. This lockdown gets the locals used to the new settlers and the new settlers used to the locals. It also keeps the newcomers inside for a few sunsets so that they understand that the coop is the place to be when night falls.

When you do let your new birds out among the resident flock, you can do a few things to help keep the peace. Make sure your coop is ready to handle the stress of the situation. Be sure to have plenty of roost space and a few places for the meek to hide if they have to. If possible, set up a second feeding station on the opposite side of the coop so that the new birds can have a chance to run to one while the honcho is eating at the other. These small favors can be all the leeway your new arrivals will need to get over the fairly inevitable hazing period.

Sam, a scrappy Ameraucana, has proven to me that a chicken can not only hang with her farmer but also become a sidekick. She often runs to me for solidarity because as the lowest-ranking bird in the coop, she is safe in my presence. No one can peck her if she's in my arms, so in my arms is where she likes to be. Her nearly featherless back proves her place at the bottom of the chicken social scene: if you're the outcast of the clan, you get pecked on a lot. So maybe it's her underdog status, and maybe I'm making too much out of a pathetic little bird. But I can't help it. She's such a sweetheart.

When I carry hay out to the sheep, Sam follows behind like a faithful dog. When I go into the coop for morning feeding or late-day egg collecting, she flies right to me, jumps onto my hand or shoulder when I pet her head, then stomps around the yoke of my shirt in search of the perfect placement for her little taloned feet. It is a splendid thing to hold a chicken you raised from a 72-hour-old hatchling, especially one that actually likes being held. You can't help but think, "Hey, I pulled this chicken thing off! She trusts me." A homesteader rite of passage.

Cock-a-Doodle-Do...or Don't

I love my boys, but let's be honest, roos aren't for everyone, especially the urban flock keeper. If your lifestyle isn't set up for them or you have finicky neighbors, prohibitive ordinances, or an urban situation, you're probably already aboard the no-rooster train.

But if you ordered straight-run chicks, the odds of getting a cockerel in with your pullets are about 50-50. And even if you ordered all females, you still might end up with an incorrectly sexed male, which you'll know for certain by six to eight weeks when he starts to crow. But for a bit of nonscientific sexing before the audio clue, you can pull a hat trick. Literally.

When your birds are three to six weeks old, place them on the kitchen floor and let them mill about for a minute or two. Then drop a hat in the middle of the floor. The hens will shrink down or run; the males usually lift their heads and look around. It's their most basic protective instinct shining through.

If you do have a rooster in with your flock, then what? Consider the pros and cons — assuming your local laws permit leaving the decision up to you, of course.

ROOSTER CONS

Roosters crow. They don't just do it at dawn (really, more like an hour before first light) but all day long. They occasionally can be aggressive and cut you with their spurs. They can be large (my Winthrop is taller than my Toulouse gander!), and having more than one can be an invitation to unregulated cockfighting. Since there is no personality test or a bark collar for roosters, they may not make the best pets. I would not recommend them for anyone in an urban, suburban, or small-children-tended flock.

ROOSTER PROS

Roosters crow. Yes, it's all day, but if you're anything like me, you love that sound and it feels like your own rural anthem. They're also great protectors of hens and allow a free-range flock to relax, knowing their man has got their backs. A rural flock of laying hens is truly happier with a cock in their midst and will usually follow him instead of scattering around your lawn. They also make adding to your flock a lot easier since, of course, having fertile eggs means the possibility of chicks. Finally, roosters are iconic in their beauty; just having one perched on a fence post here at the farm makes me want to stick a hayseed in my mouth and buy a pair of overalls.

chapter 7

From Chicks to Chickens Months 3–6

The girls have developed lots of sounds: contented singing, rambunctious cackling, and indignant complaining, to name a few.

Laying in Wait

When your chicks reach their third month of life, you should begin planning for laying. You have about two months before those first pullet marbles start appearing, but soon enough the term "nest egg" won't just refer to your 401(k) plan. So it's time to start building and installing nesting boxes for your growing birds.

Nesting Box 101

SECOND ONLY IN IMPORTANCE to your chickens' shelter are their nesting boxes. A nest box provides both a place for a hen to lay — they prefer laying in dark, out-of-the-way places — and a convenient place for you to retrieve their eggs. A nest box also helps to ensure that the eggs will remain clean and unbroken.

A good nest box design is very basic: It has one entrance, it's sheltering (hens like the protection provided by things like boxes, crates, and buckets turned on their side), and it's elevated. Also, it should have a small piece of wood for a railing across the front to keep bedding from sliding out onto the coop floor and to help the birds gain some footing when jumping up to lay.

Like nearly everything else in the chicken-husbandry world, nesting boxes can be as simple or as elaborate as you want them to be. My chicken mentor, Diana, used an ingenious system of old wooden drawers raised slightly above the bedding by the occasional cinder block. I use old wooden crates screwed into the walls of the coop, about 3 feet off the ground. Some people attach square 5-gallon plastic cat-litter buckets to their coop walls — they are dimensionally perfect for a proper nest. If that sounds somewhat odd, hey, we'll take inspiration and instruction wherever we can find them — even the most unlikely of places.

Of course, farm catalogs have great prefab wall-mounted nesting systems, but they're extremely pricey. True, investing in them means having the certainty of a proper setup for your birds, but chances are you can look at the catalog photos and realize that the plastic crates holding your old disco records can be reincarnated as nest boxes with just a few minor adjustments and a power screwdriver.

Manure Happens

Each of your hens will produce — get this — 45 pounds of poop a year. That is a lot of free fertilizer, and if you play your cards right it can be the manna from heaven your tomatoes have been begging for.

Turning your birds' coop waste into compost is a great side benefit of keeping chickens. You just need to be careful in how you use it. Since chicken manure is considered "hot," meaning very potent in nitrogen, it needs to be used sparingly in its fresh form or composted down over the course of a year before it can be turned over in your garden as a humus booster.

I use the deep-bed method for my coop. This means instead of cleaning out the old bedding and replacing it every time I freshen the coop, I pour fresh straw or pine shavings on top of the old stuff instead. This creates the necessary alchemy for the waste below to heat and compost underneath the feet of my birds. And when I clean out that coop in the spring, I have a powerhouse of nutrients for my garden.

Since most people line their coops with a bedding of wood shavings or straw, you already have the carbon ingredient ready to go, and by adding your birds' old waste and bedding into your compost turner or onto your compost pile, you are not only recycling, you're creating topsoil that will make the onions for your omelets sing for joy.

We keep our birds confined to their coop unless we are out in the yard with them, since a lot of our neighbors let their dogs run free. A drawback to confinement in a smaller space is that the more they grow, the more they devastate the lawn beneath them by eating as much green as they can and scratching away what they can't. We minimize their impact by moving our portable coop every day to a fresh patch of forage.

Chicken manure becomes a valuable fertilizer once it's composted. If you have more than you can use, gardeners will happily take it off your hands.

FINALLY!

And then one day it happens. . . . You go out to the coop on a brisk September morning — feed scoop in hand and fresh water sloshing in a bucket — thinking it's going to be like every other morning with the girls, who are cooing and clamoring for breakfast. But then something makes you tilt your head and stop in your tracks. You can just *tell* something has happened. Something feels new and exciting. As the hens all rush to chow down on their morning meal, you walk over to the nearest nest box and . . . yes! There it is, as beautiful as you imagined it would be! The first egg has finally arrived! Congratulations!

Your big girls are ready for layer rations now, and it's time to switch to a new feed.

AM I SUPPOSED TO EAT THOSE?!

After all your patient waiting, your girls have finally rewarded you for all that good care. However, those first eggs may not be what you had in mind. Sometimes they'll be so tiny, they'll look like robins' eggs. And some may not even have a yolk. Occasionally you'll get an egg laid without a shell — just a kind of white membrane holding a small yolk. Those first eggs may have specks or streaks of blood on them, too.

All of this is normal. An eighteen- to twenty-week-old pullet needs to mature into her new occupation. Hens need to work up to those jumbo eggs you're used to seeing in the store. I promise you, though, even that first little pullet egg will be worlds apart from any store-bought variety.

Honey's eggs are consistently round and fat in appearance — more of an oval than an egg shape.

Tilda's eggs usually have an uneven shell color, with little white speckles all over.

Amelia's eggs are impeccably perfect in shape and shell, but you're not likely to find eggs this color at your local market.

Chick Diary

In the industrial model of farming, uniformity of product is a key goal, so the eggs in a supermarket carton are identical in color, size, and shape. None of our chickens would pass muster as an industrial layer, but we treasure their uniqueness. It's satisfying to be able to identify which eggs belong to whom.

Even once hens are mature, you can still occasionally get an odd egg. This one from Amelia, with its mottled color and calcium deposits, looks as if it has the plague. Hens have bad days like everyone else and usually snap back to normal right away.

Fair Game

If you have done the work of raising a healthy, slick flock of chickens, you might want to strut your stuff — and theirs — at a local county fair or poultry show. The American Poultry Association, 4-H, and many local poultry clubs hold events where bird lovers can enter, chat, and talk chickens all day long. It's a chance to swell with pride when the animal you raised from a day-old chick comes home with a blue ribbon. I mean, you worked hard to bring up that spunky little hen into the beautiful Buff Orpington she is today, as gold as Jason's fleece and as bright eyed as a day-old calf. Being able to show her off and maybe win something is a lot of fun. And hey, a little recognition is nice, isn't it?

I entered my farm-raised Silkie bantams in the Bonner County Fair in Idaho, and my hen and cockerel pair took second place in their division. We didn't come home with any trophy, but the prize did cover their entry fees, and the time spent in the poultry exhibition area turned out to be a hotbed of conversation, laughs, bartering, and education. I learned so much by just chatting with other hobbyists as I fed and watered the entrants and walked around the fair looking at other chickens. It was an experience I hope to have again.

If fairs aren't your scene, you can find other ways to get active in poultry outside your own coop. Your kids can get involved in 4-H, and you can join your local breed or poultry club. The clubs usually revolve around showing, but they're also a wellspring of information and experience.

Chick Diary

When we hold a treat like clover over our girls' heads, they'll leap into the air to grab it. All of us, including the chickens, find this endlessly amusing.

Check out those drumstick gams!

In a cardboard box tucked under my right arm, a pair of Black Silkie bantams were about to make their county fair debut. It was my first fair, as well. After being instructed in how to fill out the forms and hang the cage tags, plus the care and feeding responsibilities of the exhibitors, I was ready to show my birds. Those little black chicks I had nurtured from infancy strutted and crowed for the whole county to see. I was genuinely proud of them; they were bright eyed and seemed happy.

The night before I had washed them in natural dish detergent and gently blow-dried them on towels while we listened to the back porch bluegrass show on NPR. The notion that a few years ago my New York City design school friends and I were skipping through galleries in Chelsea and eating overpriced French toast at the Empire Diner made me laugh out loud while I dried my pathetic wet hen's face with a towel. Unlike many of those old college friends, I had yet to travel overseas, but I still felt I had made one heck of a journey.

The fair lasted a whole week. I didn't know the first thing about conformation of Silkie bantams, but I ended up winning a few ribbons. I also ended up with the grand champion rooster. These kinds of things just happen with poultry people. A couple of us were in the poultry barn, squawking louder than the hens about our birds, and soon we had a trade going on. She would take home my Black Silkie rooster, and I would take her regal Welsummer.

Within minutes of bringing him home, he conquered one of my hens. If nothing else, I made one impressive chicken yenta.

Chicken Hypnosis

YOU CAN PUT A CHICKEN INTO A TRANCE that lasts anywhere from 15 seconds to 30 minutes. In addition to being a curious trick, it can be a good way to settle a bird that you need to inspect for injury or disease. Here are three methods:

- Hold her head to the ground by placing her on her side, and with a stick or finger, draw a continuous line from the tip of her beak straight outward.
- While you are cradling her belly-up in one arm, lightly massage each side of her breastbone with your fingers.
- Holding her in one arm, place her head under a wing and gently rock back and forth.

Egg-stending the Season

Chickens lay eggs when their instincts tell them that their babies have a good chance of survival, which is in the spring and summer. If you want to keep egg production up, even in midwinter, hens need to be fooled into thinking it's still reproduction season. This requires artificial lighting. For peak laying productivity, augmentation should start when daylight falls below 15 hours per day, usually in September. Your artificial lighting program needs to be maintained until the following spring, when you can let nature take over again. If you forget to turn the lights on for even one day, your hens may go into a molt and stop laying.

If you're not always home on time to flip the switch, set up a timer on your coop's light. Mine automatically flips on at 4 PM in the winter, so even if I'm late at the office, I pull into the driveway to see the henhouse light already on. If you have a timer, you also won't forget to turn the lights off. Your flock needs a good night's rest, and the girls won't sleep as well under glaring lights.

You can use either incandescent or fluorescent bulbs. Fluorescent bulbs are less expensive to run and, since they use less power, are more eco-friendly, but be sure that you get the ones labeled "soft white." Counterintuitively, fluorescent bulbs labeled "daylight," imitating the appearance of sunlight, have a cooler color temperature and will not stimulate a chicken's reproductive system.

If an electrical socket is not an option in your coop, you can set up a relatively inexpensive solar electric system like the ones commonly available for RVs. Look for a solar-charged battery powering a 12-volt light.

To keep your hens laying eggs through the winter, supplement daylight with artificial light.

The disadvantage of a lightweight, portable coop is that it doesn't offer enough protection against the very cold winters we have in the Northeast. During the coldest months, we keep our coop in the garage, which shelters the chicken run and keeps the girls warm. I lay down a heavy-duty tarp to protect the garage floor and add a deep layer of straw to absorb the manure. I muck out the coop once a week and put in fresh litter. The old, soaked straw makes premium compost.

Chick Diary

Greedy for greens now that there is a lack of forage

Heavy snow destroyed the roof of the coop. Due to the collapse of their home, the two remaining Silkies have moved inside. They spend most of their time in the garage, but I let them hang around the kitchen for about an hour every night. They drink out of the dog bowl and roost on the small metal planter of dirt that used to hold lettuce heads but now just has a coating of chicken feed on it. Which means they have soft earth to scratch in even though it's 15 degrees outside.

Last night the quiet, calm pair sat on the top of the couch to watch TV with me. Every so often they would coo or cock their head but then pretty much just fell asleep. It's kind of nice having them inside. I never had a parrot, but birds make darn good company. They'll sit on your lap and eat out of your hand.

They like to be petted and chase each other like cats or dogs would. They have never had an "accident" indoors. I don't know why some people are convinced that all animals want to do when they get indoors is defecate on a carpet — as if they themselves run to the bathroom every time they visit someone's home.

Anyway, the chickens aren't house pets. They'll be back outside for good when the thaw hits and they have a new home. But right now it's kinda fun to walk into the kitchen and see a tiny black chicken perched on the microwave preening its feathers.

chapter 8

It's Egg Time

All about Eggs

Now that your girls are finally laying, here are some important egg facts you should know.

RATE OF LAY

Most hens need 24 to 26 hours, start to finish, to lay a single egg, so at most your birds will be able to produce one egg per day. But this is a variable, depending on temperature, light, breed, and age.

The chart on pages 114 through 119 has information on the laying productivity of different breeds. At the top end of the scale, the Leghorn will produce about 300 eggs per year. Many other breeds lay far fewer. Even individuals within a single breed will lay at different rates. In a commercial farm setting, birds that lay fewer eggs than the average for their breed are culled.

Just as a hen will lay less or not lay at all when she's getting less than 15 hours of light (see page 102), production will slow down when temperatures fall below 45°F (7°C) or rise above 80° (27°C). She will also stop laying when she molts (see page 107).

A pullet will reach full productivity when she's thirty to thirty-four weeks old, and her eggs will be a normal size by that point. Hens reach their peak in the spring, and a good layer will fill your basket with five to six eggs a week. After she molts, her eggs will be larger but slightly fewer.

Once a hen turns three years old, she's past her prime, and you'll notice a considerable drop in production. She may continue to lay until she's ten or twelve, though in steadily decreasing numbers.

SHELL COLOR

Depending on the breed, eggshells can be white, cream, tan, blue, green, pink, even a brown so dark it's brick red. As a rule, hens with white earlobes (yes, they're called earlobes) lay white eggs, and hens with red earlobes lay brown eggs. Like every rule, there are exceptions to this one, and we're not even factoring in the Araucana and Ameraucana breeds, which lay those blue beauties. Some slight variations in hue aside, each hen will lay eggs of a specific color — unless she is stressed or ill (which will result in paler brown eggs) or aging (ditto).

A breed's earlobe color determines the color of its eggs.

YOLK COLOR

Egg yolks, like eggshells, can vary widely in color — but the raw yolk color depends strictly on a hen's diet, not on her breed.

RAW YOLK COLOR	CAUSE
Green	• acorns • shepherd's purse
Orange to dark yellow	• green feed • yellow corn
Reddish, olive green, black-green	• grass • cottonseed meal • silage
Yellow, dark	• alfalfa meal • marigold petals
Yellow, medium	• yellow corn
Yellow, pale	• coccidiosis (rare) • wheat (fed in place of corn) • white corn

Molting

DOGS AND CATS SHED. CHICKENS MOLT. Molting is the annual shedding of feathers from your chickens and around here takes place during the fall, when daylight shortens and the birds' biological clock tells them to refuel their plumage. During this period egg laying ceases. Good layers molt fast and furiously; they'll look like punk rockers with bald spots or weird feathers, but it's totally normal. Within two to three months they'll have a brand new set of feathers. Poor layers take longer; they really dog it, actually, taking as much as half a year to up their set.

If you want to speed up the molting process, understand that your girls' feathers are 95 percent protein. By adding some extra oomph to their diet, you can up their feather-growing production and return them sooner to their normal routines of laying and living in your coop. The extra protein can come in the form of some meat and fish along with their usual ration of layer feed. You could also add a high-quality cat food to their diet, which is higher in protein than dog food and gives the birds a little jolt of P power while they work toward their new outfits.

EGG SIZES

Naturally, the size of a hen's egg will vary with breed and age. The older the hen, the larger the egg. But according to the USDA, eggs come in six standard sizes. The sizes and their average weight are:

PEEWEE	1.25 ounces	35 grams
SMALL	1.5 ounces	43 grams
MEDIUM	1.75 ounces	50 grams
LARGE	2 ounces	57 grams
EXTRA LARGE	2.25 ounces	64 grams
JUMBO	2.5 ounces	70 grams

NUTRITIONAL VALUE

Often called the perfect food, one egg contains nearly all the nutrients necessary for life, including vitamins A, D, E, beta carotene, folic acid, and omega-3 — vitamin C is the exception. One large egg has roughly 70 calories (of which 12 percent are fat), 6.25 grams of protein (about the same as an ounce of lean meat, poultry, fish, or beans), several antioxidants (including lutein and zeaxanthin, which help prevent macular degeneration), choline, which aids brain function, and lecithin (which may also help brain function). And despite some health-food-store claims to the contrary, the color of the shell has no influence whatsoever on the nutrition contained within.

Keeping Eggs Clean — and Cleaning Dirty Eggs

Ideally, eggs should be clean when you collect them. Despite the fact that a hen's vent pops out poop as well as eggs — one bodily opening per customer — eggs come into the world quite clean thanks to the bird's oviduct, which effectively closes off the intestinal opening. (Ingenious, right?) But eggs gathered from a nest may still have dirty shells, the result of either your hens tracking mud into their nests or, yes, chicken poop landing on them after the fact.

To keep eggs from getting dirty in the first place, you'll need to remedy the source of the problem. If mud is the offender, more often than not a muddy entryway to the coop is the culprit, and the nesting boxes are too close to the ground. Raising the boxes should help. If bird droppings are befouling the eggs, it's likely that the rails the hens roost on are too close to the nest itself, causing their droppings to land smack dab on the eggs. Alleviate this problem by making sure the rail is no closer than 8 inches from the nest edge.

If you do find a particularly foul-looking egg, soiled by either mud or excrement, toss it. That egg will be covered with bacteria and not safe to eat. A slightly dirty egg, on the other hand, can be brushed off or rubbed with either a sanding sponge or a nylon scouring pad. Water should be avoided since washing the shell rinses away its natural seal against bacteria. If you do wash an egg, use water that's slightly warmer than the egg itself. If the water is cooler than the egg, bacteria can actually be drawn through the shell into the egg. Dry the cleaned egg before putting it into a carton and use it as soon as possible.

EGG SAFETY

With your girls finally providing you with an egg bounty, you should strictly follow some safety basics:

- Keep nests clean and lined with fresh litter.
- Collect eggs often, and refrigerate promptly. Your flock won't necessarily keep to a tidy schedule, so check several times a day if possible.
- Toss seriously soiled and broken, leaking eggs. (However, an egg that has cracked slightly is safe to eat as along as the membrane that's attached to the shell is still intact and the egg is used right away.)
- Clean moderately dirty eggs.
- Cook eggs or egg-rich foods to 160°F (70°C), and either serve immediately or refrigerate. Although salmonella poisoning is unlikely from homegrown eggs, cooking them will destroy any salmonella bacteria that may be present.
- Wash hands and cooking utensils after handling raw eggs.

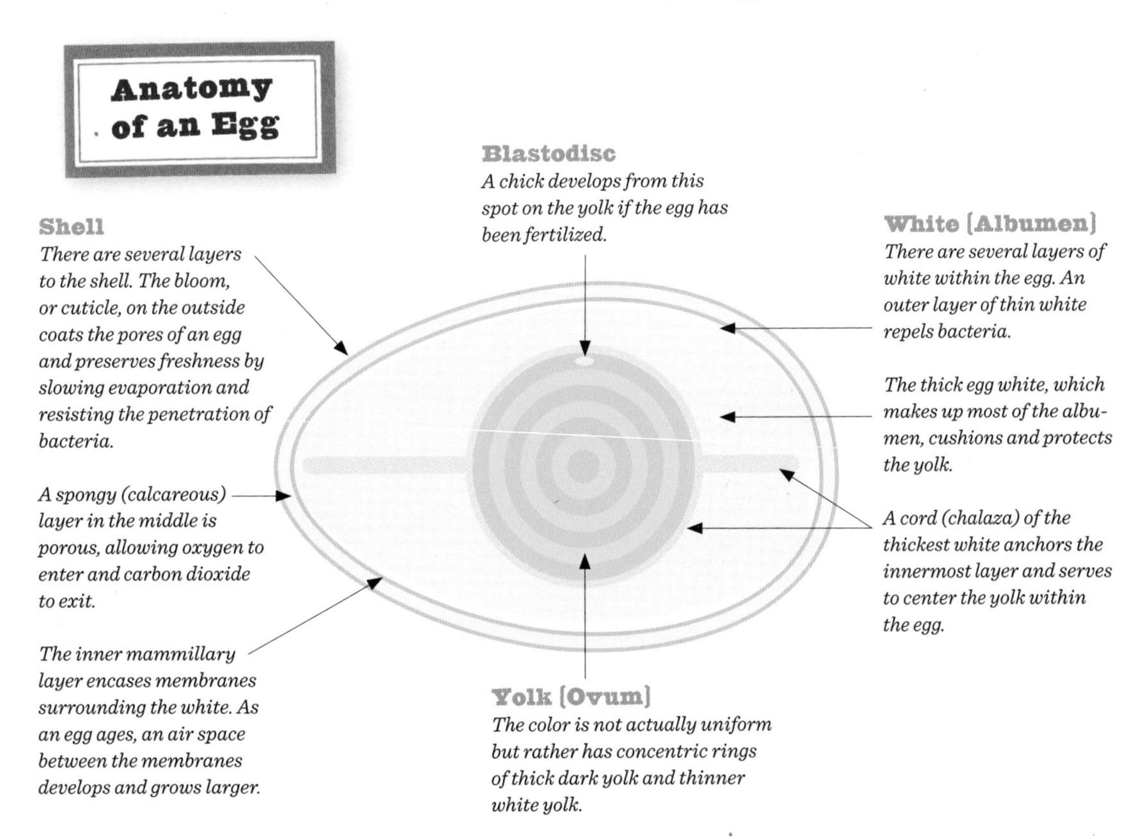

Broodiness

Springtime can trigger a chicken's instinct to incubate, and when a hen gets into parenting mode, she becomes what's called "broody." *Broody* is a term that means the gal has set on a nest and doesn't plan on moving until what's under her hatches. Some flock owners regard this as a trait to be coveted and respected. Broody hens bring chicks into the world, after all, and what better way to increase your flock than with a volunteer grain-fed incubator. Other folks feel a broody hen is a nuisance. Since a hen that sets is no longer laying, she stops producing and gets in the way of easily collecting eggs for use or sale.

If you want to discourage the hatching impulse, repeatedly remove a broody hen from the nest and make sure no eggs accumulate in her spot. Often the absence of eggs, plus being placed back on the ground or outside the coop, is enough to kick the PTA-mom habit. You can also try moving or covering her nesting site.

If one of your hens is still dead set on hatching a clutch, your choice is either to let the broodiness run its course (which will last two to four weeks), or to break her habit in a broody coop.

A broody coop is confined housing without any nesting space, usually a wire cage raised off the floor. You can break most hard cases in one to three days, though it may take her a week or longer to begin laying again.

It may seem extreme, but a hen that isn't laying isn't earning her keep for most backyard farmers, so the broody coop could be the remedy. The small-flock raiser would do well to have just such a cage anyway in case a bird is injured or needs to be separated, since a broody coop could also double as a hospital unit or therapy center.

A broody hen will bristle her feathers when you disturb her on the nest, make threatening noises, and peck your hand if you get close.

Chick Diary

Honey got broody at the end of winter. As a last resort, we got a dog crate to make a broody coop. We raised it off the floor on two-by-fours so she couldn't hunker down to nest. We made a roost out of another two-by-four, with the corners cut at an angle and then sanded so she'd be comfortable. We put screws in at the board's ends so it could hook onto the sides of the crate.

If I had to do it over again, I would have let the broodiness run its course. We don't consider ourselves farmers. Our chickens are pets. Honey's removal and isolation seemed stressful not only for her but for the rest of the flock.

THREE-HEN QUICHE

OVER MY GIRLS' COOP, A SIGN READS: TEAM QUICHE. *That's because each of my three original hens would lay an egg a day, and three eggs are exactly what my favorite recipe calls for. Quiche is easy to make, great to share, and keeps well in the fridge. It tastes just as good heated up in the oven the next day. You can customize it, adding whatever you prefer. If your add-ins include breakfast meats such as bacon or sausage, make sure they're precooked before mixing them with the other ingredients.*

ingredients

- Olive oil
- ⅓ cup diced onion
- ½ cup water
- 1 broccoli crown, chopped into inch-long pieces
- 3 large eggs
- 1 cup milk
- Salt and pepper to taste
- ½ to 1 cup grated cheddar and Jack cheeses
- 1 piecrust (frozen or homemade)

Preheat the oven to 350°F (175°C). Heat the olive oil in a skillet on medium-high heat. Sauté the onion until browned and fragrant; set aside in a small bowl. Pour the water into the skillet, and steam the broccoli over medium-high heat. Sprinkle a little salt over them while they steam. When the broccoli is tender, set it aside in a small bowl.

Beat the eggs together with the milk. Add a little salt and pepper to taste. Toss the onion and broccoli together and spread over the piecrust (in pie pan). Pour the egg mixture over the vegetables. Sprinkle the cheeses over the top. Bake uncovered for 45 to 50 minutes or until a knife tip poked into the center comes out clean. You've now got yourself a quiche.

Spreading Your Wings

You've followed us on the path from chick to egg. Page by page, you've watched our girls Tilda, Amelia, and Honey grow from little firecrackers into feathered fireworks. It's been a nice country stroll, but now that they've matured, the pace will pick up.

Once the eggs start coming, the possibilities from these birds are endless. Everything from the best baked goods you've ever eaten to starting your own small egg business can launch from here. Maybe you're now considering having 20 birds laying nearly two dozen eggs a day, and you'll be printing out flyers to sell the eggs at your office. Or maybe you've simply mastered the perfect ratio of egg to milk in your children's Saturday morning French toast. It doesn't matter how far you spread your wings; just making the effort to bring some good homegrown food to the table is enough.

Again, I congratulate you, but getting pullets to a regular laying schedule can be just the beginning of the story. From here you can take the lessons and skills you've learned and expand them to further poultry adventures that include ducks, geese, and turkeys. Even their brooder equipment is the same. Or perhaps chickens will get you in the right mental state for husbandry and be your springboard to other farm animals, such as goats and sheep. Consider chickens backyard farming 101.

Consider that first egg your diploma. In case I haven't said it enough yet: Welcome to the coop.

BREED	ALBC CONSERVATION PRIORITY RANKING (2010)	USE	EGGSHELL COLOR	EGG SIZE	HEN SIZE
AMERAUCANA	Not listed	Eggs, Meat	Green, blue	Med.-Lg.	Medium
ANCONA	Watch	Eggs	White	Med.-Lg.	Medium
ARAUCANA	Study	Eggs, Meat	Blue	Med.-Lg.	Small
AUSTRALORP	Recovering	Eggs, Meat	Brown	Lg.	Heavy
BARNEVELDER	Not listed	Eggs, Meat	Dark brown	Med.-Lg.	Medium
BLUE ANDALUSIAN	Threatened	Eggs	Chalk white	Lg.	Medium
BRAHMA	Watch	Eggs, Meat	Brown	Med.-Lg.	Very Heavy
BUTTERCUP	Threatened	Eggs	White	Sm.-Med.	Small
CAMPINE	Critical	Eggs	White	Med.-Lg.	Small
CATALANA	Watch	Eggs, Meat	White to Tinted	Med.	Medium
CHANTECLER	Critical	Eggs, Meat	Brown	Lg.	Heavy
COCHIN	Watch	Meat, Ornamental	Brown	Med.-Lg.	Very Heavy

Pick a breed, any breed! This chart will help you shop for the perfect girls to start or add to your flock. Aside from listing birds and their characteristics, ranging from weather tolerance to eggshell color, it also shows their status as determined by the American Livestock Breeds Conservancy (ALBC), which is actively trying to keep breeding populations of heritage breeds thriving. Consult the oracle. You might well find a rare, high-producing, cold-weather-loving, nonbroody breed that's perfect for your Chicago frittata flock.

RATE OF LAY PER YEAR	BROODY	BEARS CONFINEMENT	CLIMATE TOLERANCE	SPECIAL CHARACTERISTICS
150	No	Yes	Cold-hardy	Egg color / Muff and beard / Calm and docile
120–180	No	Will tolerate	Cold-hardy	Very active and avoids human contact / Noted for hardiness and vigor / Good forager
150	Yes	Will tolerate	Cold-hardy	Egg color / No tail / Tufted head feathers
250	Yes	Yes	Cold-hardy	Productive and fast growing / Active, yet very docile and friendly
150–200	No	Yes	Cold-hardy	Egg color / Active, yet docile and friendly
150	No	No	Heat-tolerant	Very rugged, robust, and active / Excels in free-range conditions
140+	Yes	Will tolerate	Cold-hardy Heat-tolerant	Meat quality / Winter layer / Calm and docile / Good setters and mothers / Slow to mature
100	No	No	Heat-tolerant	Unique buttercup-shaped comb / Very active / Excels in free-range conditions / Good forager
150	No	Will tolerate	Cold-hardy	Good forager / Very active and smart / Must protect large combs from frostbite
200	No	No	Heat-tolerant	Meat quality / Good forager / Very active
120–180	No	Yes	Cold-hardy	Can be eaten at any age / Winter layer / Calm and docile
100–140	Yes	Yes	Cold-hardy	One of the best setters, especially for turkey and duck eggs / Slow to mature / Calm, friendly, and docile

BREED	ALBC CONSERVATION PRIORITY RANKING (2010)	USE	EGGSHELL COLOR	EGG SIZE	HEN SIZE
CORNISH	Common	Meat	Light brown	Sm.	Very Heavy
CREVECOUR	Critical	Ornamental	White	Med.-Lg.	Heavy
CUBALAYA	Threatened	Eggs, Meat	White	Sm.-Med.	Medium
DELAWARE	Critical	Eggs, Meat	Brown	Lg.-XLg.	Heavy
DOMINIQUE	Watch	Eggs, Meat	Brown	Med.-Lg.	Medium
DORKING	Threatened	Eggs, Meat	White	Med.-Lg.	Heavy
EGYPTIAN FAYOUMI	Not listed	Eggs	Tinted	Sm.	Small
FAVEROLLE	Threatened	Eggs, Meat	Tinted to light brown	Med.-Lg.	Heavy
HAMBURG	Watch	Eggs	White	Sm.-Med.	Small
HOLLAND	Critical	Eggs, Meat	White	Lg.	Medium
HOUDAN	Watch	Ornamental	White	Sm.-Med.	Very Heavy
JAVA	Threatened	Eggs, Meat	Brown	Med.-Lg.	Heavy
JERSEY GIANT	Watch	Eggs, Meat	Brown	XLg.	Very Heavy
LA FLECHE	Watch	Eggs, Meat	White	Med.-Lg.	Medium
LAKENVELDER	Threatened	Eggs	White to tinted	Med.	Small
LANGSHAN	Threatened	Eggs, Meat	Very dark brown	Med.-Lg.	Heavy

	RATE OF LAY PER YEAR	BROODY	BEARS CONFINEMENT	CLIMATE TOLERANCE	SPECIAL CHARACTERISTICS
	50	Yes	Yes	Cold-hardy	Meat quality / Loud and active / Not suitable for free range
	100	No	Yes	Moderate	Very active and require exercise / Crested head feathers
	125–175	Yes	Will tolerate	Heat-tolerant	Meat quality / Fancy tail feathers / Can be aggressive and noisy but mild-mannered compared to other game birds
	200	Yes	Yes	Cold-hardy Heat-tolerant	Quick to mature / Can be eaten at any age / Calm, friendly, and docile
	230–275	Yes	Yes	Cold-hardy	Good forager / Good setters and mothers / Calm and docile
	150	Yes	Yes	Cold-hardy	Winter layer / Good forager / Red earlobes but white eggs / Five toes / Calm and docile
	100	Yes	No	Heat-tolerant	Quick to mature / Good forager / Very active, vocal, and flighty / Upright tails
	200	Yes	Yes	Cold-hardy	Calm and docile / Muff and beard, feathered feet, and five toes / Winter layer / Roosters less aggressive than most
	200	No	No	Cold-hardy	Very active and flighty / Excels in free-range conditions
	150	Yes	Yes	Cold-hardy	Good forager
	100	Yes	Yes	Moderate	Setters, but tend to break eggs / Calm and docile / Crested head feathers, beard, and five toes
	100	Yes	Yes	Moderate	Good forager / Calm and docile
	150	No	Yes	Cold-hardy	Unusually large size / Meat quality / Slow to mature / Calm and docile
	150	No	Yes	Moderate	Good forager / Meat quality / Slow to mature / Very active and avoid human contact / V-comb
	150	No	Yes	Moderate	Good forager / Active
	150	Yes	Yes	Cold-hardy	Slow to mature / Calm and docile

BREED	ALBC CONSERVATION PRIORITY RANKING (2010)	USE	EGGSHELL COLOR	EGG SIZE	HEN SIZE
LEGHORN	Common (White); Recovering (non-White)	Eggs	White	Lg.	Medium
MALAY	Threatened	Meat	Brown	Med.	Very Heavy
MARAN	Not listed	Eggs, Meat	Dark brown	Lg.	Heavy
MINORCA	Watch	Eggs	Chalk white	Lg.-XLg.	Heavy
NAKED NECK	Common	Eggs, Meat	Light brown	Med.	Medium
NEW HAMPSHIRE	Watch	Eggs, Meat	Brown	Lg.-XLg.	Heavy
ORPINGTON	Recovering	Eggs, Meat	Brown	Lg.-XLg.	Heavy
PENEDESENCA	Not listed	Eggs	Dark brown	Med.	Small
PHOENIX	Common	Ornamental	Cream or tinted	Sm.-Med.	Small
PLYMOUTH ROCK	Recovering	Eggs, Meat	Brown	Lg.	Heavy
POLISH	Watch	Eggs, Ornamental	White	Med.-Lg.	Small
RHODE ISLAND	Recovering (Red); Watch (White)	Eggs, Meat	Brown	Lg.-XLg.	Heavy
SEBRIGHT BANTAM	Watch	Ornamental	White or creamy	Peewee	Bantam
SILKIE BANTAM	Common	Ornamental	Cream or tinted	Peewee	Bantam
SPANISH	Critical	Eggs	Chalk white	Lg.	Medium
SUSSEX	Recovering	Eggs, Meat	Tan to brown	Lg.	Heavy
WELSUMMER	Not listed	Eggs, Meat	Deep terra-cotta	Med.-Lg.	Medium
WYANDOTTE	Recovering	Eggs, Meat	Brown	Lg.	Heavy
YOKOHAMA	Common	Ornamental	Cream or tinted	Sm.-Med.	Small

RATE OF LAY PER YEAR	BROODY	BEARS CONFINEMENT	CLIMATE TOLERANCE	SPECIAL CHARACTERISTICS
300 (White)	No	Yes	Cold-hardy (White only); heat-tolerant	Excellent layer / Very active / Noted for hardiness and vigor
50	Yes	No	Heat-tolerant	Tallest of all chickens / Aggressive / Slow to mature
150-200	No	Yes	Moderate	Egg color / Active
200	No	Will tolerate	Heat-tolerant	Good forager / Very active
100	Yes	Yes	Cold-hardy Heat-tolerant	Calm and especially easy to tame / Featherless neck
150	Yes	Yes	Cold-hardy Heat-tolerant	Quick to mature / Calm and docile
150	Yes	Yes	Cold-hardy	Meat quality / Calm and docile
160	No	No	Heat-tolerant	Egg color / Active and avoid human contact / Carnation comb
50	No	Yes	Moderate	Fancy tail feathers
200	Yes	Yes	Cold-hardy	Quick to mature / Meat quality / Excel in free-range conditions / Docile
100-200	No	Yes	Moderate	Crested head feathers / Calm and docile / Rate of lay unreliable within breed
250	No	Yes	Cold-hardy Heat-tolerant	Meat quality / Active yet docile
50	No	Will tolerate	Heat-tolerant	Cocky but not aggressive
150	Yes	Yes	Cold-hardy Heat-tolerant	Excellent setters and mothers / Calm and docile / Hairlike plumage, black skin and bones, and five toes
150	No	Will tolerate	Heat-tolerant	Active and noisy / White faces
200	Yes	Yes	Cold-hardy	Meat quality / Winter layer / Calm and docile
150-200	Yes	Yes	Cold-hardy	Egg color / Good forager / Intelligent and friendly
200	Yes	Yes	Cold-hardy	Calm and docile / Tend to be dominant
50	No	Yes	Moderate	Much the same as the Phoenix but has a walnut comb instead of a single comb

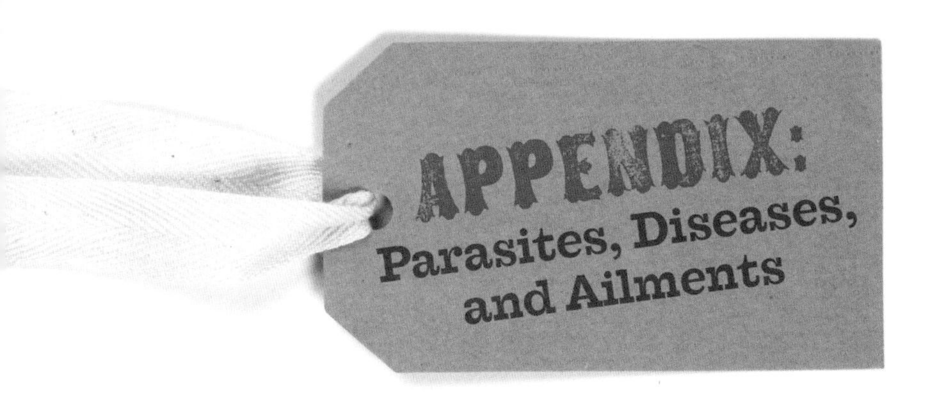

Chickens are not fragile animals. A well-maintained flock in clean bedding may never even sneeze, but sometimes the healthiest flock can pick up a parasite or catch a disease from outside your control. Here are a few common problems observed in backyard flocks. If you notice a drooping head, sneezing, or general weakness it's best to do some research or call your local extension to address the problem straight away.

AVIAN INFLUENZA

Also known as bird flu, avian influenza gave us all a scare a few years ago when the disease was transmitted from birds to people in Asia. While it made headlines, the actual ailment is much less riveting. Avian influenza is a rare disease, scarcely a worry to people with small backyard flocks and clean coops. Despite the hype, unless you are sharing a bedroom with your chickens in a dirty cage: Don't lose any sleep over it.

BRONCHITIS

Just like us, chickens can get colds and some of them aren't pretty. A bout of raspy bronchitis can sweep through a flock, causing the birds to breathe heavily (especially when roosting and trying to sleep at night), sneeze, discharge through the beak and eyes, and experience general malaise. It's not life threatening, but it can ruin egg production for a few weeks. There is no cure-all, but offering a tablespoon of apple cider vinegar or Pedialyte per gallon of drinking water can help a lot. Electrolytes and clean fluid are what they need. When my birds come down with infectious bronchitis (as they do in the cold months every few years) I buy them a new font, keep it clean, and add the vinegar. They always bounce back, but you'll miss the omelets while they're ill. Bottom line: Proper ventilation, clean, dry bedding, and adequate space will go a long way toward preventing colds in chickens.

BUMBLEFOOT

A common issue in the henhouse, bumblefoot is most often found in males of heavy breeds but can affect laying hens as well. It features a large, gross, pus-filled abscess on the pad of the chicken's foot, an infection that can eventually kill the bird if it is left untreated and spreads. The best prevention is a clean coop. If it's too late and you have a lame hen with a sore, bleeding foot you must capture her, restrain her, clean her foot, and soak it in warm water and Epsom salts. (Make sure to wear latex gloves!) Then spray on Neosporin, wrap the leg, and let the bird heal in a separate, safe, clean cage. Repeat many times.

CALCIUM DEFICIENCY

You might walk into the coop one day and find the ghost of an egg in the nest box: a filmy, shell-less blob. If the weather is hot and your birds seem to be drooping, they might have calcium deficiency. A hen's eggshell is 94 percent calcium carbonate so, as you can imagine, your birds can use a vital boost if you can supply it. I feed my hens oyster shell in a small pie pan by their feed. Others offer a crushed limestone, or dry out and crumble the hens' old shells and add them back to their feed. Make sure the crushed shells do not resemble eggs, lest your birds get ideas about snacking on their own.

COCCIDIOSIS

Caused by a dangerous parasite, this brutal intestinal disease is often associated with younger birds — and chick loss. It removes the birds' interest in life. They no longer care to drink or eat and become droopy and listless. This easily transferable and high-mortality wasting disease is the best argument for keeping babes on medicated feed. It is spread through droppings, so a clean brooder and coop are the best prevention. Most coccidiostats (drugs that control coccidiosis) are sulfa-based antibiotics. Antibiotics can treat it (survivors make it back in 10–14 days), but a clean environment and medicated feed beat the stress and expense of treating your flock, many of which you may lose.

FOWL POX

Chickens can come down with chicken pox too, but it's not related to the red, itchy bumps we might remember from childhood. Fowl pox is a wartlike infection that leaves scabby growths on the birds' non-feathered areas (like the head, legs, feet, and vent). Sometimes it infects the inside of the animal instead causing growths in the mouth or airways. It slows the animals down and can halt egg production. It can be transferred from bird to bird through open wounds or mosquitoes. The good news is that fowl pox is slow to spread, so if you find an affected bird you can remove it from the flock and the others may not catch it. Let the hen heal on its own (which may take two weeks) and return it when the warts are gone and the bird appears healthy. There is no known treatment. It cannot be spread to humans.

KINKY BACK

If you are considering meat birds, this is a fairly common disease of broiler flocks, a sad side effect of engineering animals to grow faster and fatter than they should. The rapidly growing spine twists, arching or bending to the point of paralysis and crippling the bird indefinitely. Ultimately the legs give out under the massive weight. It's not pretty and there is no cure: these birds must be culled for their own sake. They will not be able to survive to harvest weight, and their lives would be sorry ones.

LARYNGOTRACHEITIS

This throat disease is caused by birds being in contact with infected droppings or with the carcasses of infected birds. Its symptoms are trouble swallowing, ruffled feathers around the neck, and watery eyes. Your hens might gasp, stretch out their necks, and suffer just to breathe. Their throats are getting clogged up from mucus, and it could kill the bird if not seen to. The virus can be treated only via vaccine, since antibiotics are useless.

LICE

Lice are not fun. These little, flat, straw-colored bugs can live on your chickens in the land between the skin and feathers and cause mild problems (like itching) or more awful ones (like so much itching your poor hens can't sleep). Chickens must be able to clean themselves, so regular free-range flocks that can ruffle their feathers, preen, and take dust baths usually keep down their louse issues through normal care. But birds with sliced-off beaks in confinement with no opportunity to gussy up may be more susceptible. If your birds get lice you can remedy it with a powdering of anti-lice-mite treatment.

MAREK'S

This unfortunate, serious, and fast-acting disease will kill a whole flock if left to spread. Marek's is a form of bird herpes that causes everything from splayed-leg crippling to spinal cysts. Symptoms include young chicks eating constantly but losing weight, white tumors with brown scabs on the skin, and paralyzed legs (one leg pointing forward and the other back). You'll notice the pupils of their eyes shrinking to pinpoint specks, along with general depression. The only prevention is vaccination, and there is no cure. Remove the bird and put it down.

WORMS

Internal parasites are common in outdoor birds — and some say that all free-range farm birds may carry at least a small amount — but too many can make a bird sick or even kill her. There are medications to clear it up, but you may not realize that parasites (being internal) are the issue until it's too late. To avoid an outbreak: Prevent overcrowding, keep birds on clean litter and bedding, and make sure the droppings of wild birds (like pigeons and songbirds) don't get mixed into the chicken's main living area.

I lost two hens. It seems that when the weather really starts to change, when the first truly cold or warm nights hit in late fall or early summer, I lose some birds. This morning a three-year-old and a three-month-old were both belly up in the coop — and another hen is starting to droop just like the other two had. I hope she kicks back into shape.

Because they're calling for snow showers I had a lot of farm prep to do in case the morning met me with a layer of powder. After a hard day's work, my final chore was to retrieve the last pumpkin from the garden, a behemoth wider than two volleyballs. I trudged out to the garden and sliced the vines with my knife, then heaved the beast over my shoulder, breathing heavily.

As I walked through the garden gate, I spied the dead little brown hen I had placed there earlier that morning. I sighed. Putting down the giant pumpkin, I carried her softly over to the compost and set her among the graceful decline. I'd raised that bird and eaten her eggs, and she had served this farm well. She deserved a few moments and a proper spot in the quiet of the pile. Now she'd become next year's vegetables. I said a hushed thank-you, heaved the pumpkin back over my shoulder, and went inside.

Online Chicken Community

A thriving online community for and about chicken owners and future chicken owners alike can be found in the form of forums, resources, and more.

AMERICAN POULTRY ASSOCIATION

www.amerpoultryassn.com

The official chicken-owners' club, the American Poultry Association is a national organization for the exhibition and breeding of countless types of chickens. If you want to hang with the big kids, this is the place to be. An APA poultry show is an amazing way to spend a Saturday. You'll get to see some of the best specimens from hundreds of breeds and varieties, swap business cards, and maybe even walk out with a purchased pullet tucked under your coat.

BACKYARDCHICKENS.COM

www.backyardchickens.com

This expansive Web site dedicated to chicken owners is waiting for you to join the forums or check out the illustrious henhouse tours. It has amazing resources and free information to download — everything from coop plans to scrambled egg recipes. The forums are lively and the mood fun. This is also the only place I know where you can get a bumper sticker that proclaims "My Pet Makes Me Breakfast."

COLD ANTLER FARM

http://coldantlerfarm.blogspot.com

This is my personal blog — the source of many of the journal entries you read here. If you want to keep up with my flock or read about the latest farm antics day by day, grab a cup of coffee, log onto the site, and enjoy the show.

FEATHERSITE

http://feathersite.com

With plenty of photos and breed information, this online reference of countless breeds and varieties of chickens and other fowl is a great place to do your homework when picking out hens for your climate and conditions. You'll find lots of encyclopedic information on those birds you dog-eared in your hatchery catalog.

THE LIVESTOCK CONSERVANCY

(formerly The American Livestock Breeds Conservancy)

www.livestockconservancy.com

The Livestock Conservancy is a collection of hobbyists and breeders of heritage livestock and poultry. These are small-farm soldiers toiling on the front lines to breed and keep animals suited for traditional fare and nonconventional farms. Which is to say animals not meant to live in feedlots but to live and produce on smaller farms. Here you can find what breeds of chickens need saving and perhaps offer to add a few to your flock. As someone who's raised some members of the conservation list, I can personally state that this organization's resources and membership are very helpful.

URBAN-HOMESTEADING.COM

http://urban-homesteading.com

Despite the name, this online resource for all things self-reliant specializes in homesteading for both city *and* suburban dwellers.

Hatcheries and Chicken Supplies

Many local farm and feed shops sponsor "chick days" when they have bins full of birds for you to choose from. These events are usually early in the spring.

Some local feed stores will special-order chicks for you. Since most hatcheries require you to order quantities that are on the high side for the backyard flock, the advantage of going through a store is that they'll put all customer orders in at once to meet those minimums. While shopping at your neighborhood store, you'll meet someone face-to-face who may have experience with chickens and can help hold your hand through the experience.

There are also excellent online and catalog sources for everything you'll need for your little farm, including the birds themselves!

LEHMAN'S

888-438-5346

www.lehmans.com

The best nonelectric catalog out there, this Ohio-based company supplies many Amish homes across America as well as modern homesteaders. From garden gear to hand-cranked washing machines, these guys have it all if you want to go back in time.

MURRAY MCMURRAY HATCHERY

800-456-3280

www.mcmurrayhatchery.com

A long-standing reputable resource, this well-loved hatchery in Iowa carries a wide variety of bantams and standard breeds as well as many heritage, rare, and fancy chickens. Minimum order of 25 birds, but they're fine with you mixing it up with geese, turkeys, and other fowl!

MY PET CHICKEN

888-460-1529

www.mypetchicken.com

A boutique operation that will ship as few as three birds to your front door (or post office), My Pet Chicken is a powerhouse of a site, selling everything from coops to feed to fancy bantams. They also have great free downloadable chicken-care guides and chicken-related gifts. It's one of the few places where you can buy both a rooster and a toaster that cooks eggs.

OMLET

http://omlet.us

Makers of the Eglu, the coolest backyard coop available to the public today. A little pricey, they'll run you about the same as a new iPod (or two), but I assure you they are much more fun.

Books and Reference

If you want to hatch chicks from eggs, butcher a bird for the table, or otherwise go deeper into chicken husbandry, here are some great reference sources. Or maybe chickens are just the start and you want to expand your homestead in other areas. We can point you in some good directions there, too.

Butters, MaryJane. *MaryJane's Ideabook, Cookbook, Lifebook: For the Farmgirl in All of Us.* New York: Random House, 2005.
The queen bee herself, Mary Jane has put together a gorgeous idea book (cookbook, life book) full of photography, musings, plans, suggestions, recipes, stories, and, of course, ideas.

Damerow, Gail. *The Chicken Health Handbook.* North Adams, MA: Storey Publishing, 1994.
A comprehensive reference covering the problems and diseases common to chickens of all breeds and ages. Practical charts identify common symptoms and causes of disease, while an alphabetical listing of diseases provides advice on treatment. A must-have on the bookshelf.

Damerow, Gail. ***Storey's Guide to Raising Chickens,*** 3rd ed. North Adams, MA: Storey Publishing, 2010.
The author has been captivated by chickens for 40 years, and her knowledge is incomparable. Her essential reference book will guide you through every chicken situation imaginable — and then some.

Ekarius, Carol. ***Storey's Illustrated Guide to Poultry Breeds.*** North Adams, MA: Storey Publishing, 2007.
More than just gorgeous photos of dozens of amazing breeds, this book is the ultimate primer for backyard farmers and chicken fanciers alike.

Emery, Carla. ***The Encyclopedia of Country Living.*** 10th ed. Seattle: Sasquatch Books, 2008.
Every home, country or not, should have one of these books on hand. The biggest, most comprehensive encyclopedia on general homesteading ever written, it has gone through a bunch of revisions, and the newest version includes Web sites and e-mail addresses.

Kilarski, Barbara. ***Keep Chickens!*** North Adams, MA: Storey Publishing, 2003.
A must-have for any urban flock owner, this was written by a Portland lady who invited a few hens into her life — and it all went uphill from there.

Madigan, Carleen. ***The Backyard Homestead.*** North Adams, MA: Storey Publishing, 2009.
This book is nothing but trouble. You'll read the chicken section and suddenly start looking at your backyard to see if you have enough space to plant barley and hops to grow your own beer. It's a handy guide that covers everything from livestock to sourdough bread — and all from a regular suburban house with a half acre or even less.

Pangman, Judy. ***Chicken Coops.*** North Adams, MA: Storey Publishing, 2006.
If you'd like to build your own coop, here are plans, elevation drawings, and basic building ideas for 45 original coops — from strictly practical to flights of fancy.

Rossier, Jay. ***Living with Chickens.*** Guilford, CT: Lyons Press, 2004.
An amazing book and resource for the new bird lover — a beautifully photographed guide and beginner's resource.

Woginrich, Jenna, ***Made from Scratch.*** North Adams, MA: Storey Publishing, 2009.
The author's story of finding her own first flock, along with other homesteading adventures involving gardening, rabbits, and honeybees.

***HOBBY FARMS* MAGAZINE**
I-5 Publishing, LLC
800-627-6157
www.hobbyfarms.com
One of the reasons I got into this mess in the first place was from thumbing through issues of *Hobby Farms* in college. That bolted me into my hunt for my own little farm. A great all-around resource.

MOTHER EARTH NEWS
Ogden Publications, Inc.
800-234-3368
www.motherearthnews.com
Forty plus years of helping people learn to live better and to live off the land. Everything from gardening tips to recipes, from how to pick out a tractor to breaking environmental news. The best investment in a magazine you can make as a small-scale homesteader.

Buying Notes

The hen trio in this book lives in an Eglu. For details, see Omlet in the Resources listings. The coop featured on page 32 is available prebuilt on Etsy at the ChickenCoops shop; you can also purchase the plans to build a design like this yourself at catawbacoops.com. The stealth coop featured on page 33 is available from Egganic Industries at henspa.com.

Page numbers in *italics* indicate illustrations and photographs, and page numbers in **boldface** indicate charts and tables.